Managing IoT Systems for Institutions and Cities

Internal Audit and IT Audit

Series Editor:
Dan Swanson, Dan Swanson and Associates, Ltd., Winnipeg, Manitoba, Canada

The *Internal Audit and IT Audit* series publishes leading-edge books on critical subjects facing audit executives as well as internal and IT audit practitioners. Key topics include Audit Leadership, Cybersecurity, Strategic Risk Management, Auditing Various IT Activities and Processes, Audit Management, and Operational Auditing.

Internet of Things, for Things, and by Things
Abhik Chaudhuri

Supply Chain Risk Management: Applying Secure Acquisition Principles to Ensure a Trusted Technology Product
Ken Sigler, Dan Shoemaker, and Anne Kohnke

Why CISOs Fail: The Missing Link in Security Management—and How to Fix It
Barak Engel

Practitioner's Guide to Business Impact Analysis
Priti Sikdar

Data Analytics for Internal Auditors
Richard E. Cascarino

The CISO Journey: Life Lessons and Concepts to Accelerate Your Professional Development
Eugene M Fredriksen

Implementing Cybersecurity: A Guide to the National Institute of Standards and Technology Risk Management Framework
Anne Kohnke, Ken Sigler, and Dan Shoemaker

Cognitive Hack: The New Battleground in Cybersecurity ... the Human Mind
James Bone

For more information about this series, please visit https://www.crcpress.com/Internal-Audit-and-IT-Audit/book-series/CRCINTAUDITA

Managing IoT Systems for Institutions and Cities

Chuck Benson

CRC Press

Taylor & Francis Group

Boca Raton London New York

CRC Press is an imprint of the
Taylor & Francis Group, an **informa** business

AN AUERBACH BOOK

CRC Press
Taylor & Francis Group
6000 Broken Sound Parkway NW, Suite 300
Boca Raton, FL 33487-2742

© 2019 by Taylor & Francis Group, LLC
CRC Press is an imprint of Taylor & Francis Group, an Informa business

No claim to original U.S. Government works

Printed on acid-free paper

International Standard Book Number-13: 978-1-138-59048-9 (Hardback)

Visit the Taylor & Francis Web site at
http://www.taylorandfrancis.com

and the CRC Press Web site at
http://www.crcpress.com

For Alex and Jackie

Contents

Author

Chuck Benson is the Director of IoT Risk Mitigation Strategy at the University of Washington. He has been at the University for over 15 years with experience in facilities IT, healthcare IT, and central IT. He has testified before the US–China Economic & Security Review Commission on IoT risk mitigation for institutions and cities, chaired university task forces on IoT systems risk mitigation, and has provided national leadership by chairing a national task force on IoT systems risk management for Internet2 for Higher Education institutions across the country. Chuck has written numerous articles, has been interviewed for articles, has blogged for several years on the topic at longtailrisk.com, and is a contributing author on IoT systems in the book, *Creating, Analysing, and Sustaining Smart Cities – A Systems Perspective.* He is also a member of the University of Washington's drone policy working group, has an electrical engineering degree from Vanderbilt University, and an MS computer science degree from Eastern Washington University. He is also a former Marine Corps Captain and helicopter pilot.

Chapter 1

IoT Systems Introduction

Internet of Things (IoT) systems are rapidly changing the world around us and will continue to do so. These systems offer substantial potential benefit in terms of social value and business value in the institutions and cities in which they are deployed. However, *how* IoT systems are selected, implemented, and operated – and the ease or difficulty of implementation and operation – has significant consequences for the success of IoT systems implementation in institutions and cities.

All systems, whether natural, technical, or social, experience systems loss. IoT systems, these sociotechnical systems deployed and geographically distributed throughout the environments around us in our cities and institutions can have elements of all three systems – natural, social, and technical. The technical part, the technology part, of IoT systems is only a fraction of the overall system. Systems losses occur within each of these different types of systems and between these systems. All of the resources that appear to be available to implement and operate the IoT system do not get converted into actual positive, value-added output. A considerable portion of these ostensible resources – staffing, time, funding, technologies, and others – simply get converted to waste. As we will see, these systemic losses involving IoT systems will manifest themselves in lost Return on Investment (ROI) and degraded cybersecurity capabilities and cyber risk profiles for institutions and cities.

This systems waste, this loss of resources, is felt particularly fully in resource-constrained environments – which almost all institutions and cities are. This waste, and thus this impact on IoT systems success, can be mitigated substantially by paying attention to the *manageability* of the IoT system. This manageability is not just technical aspects. It presents itself in the deployed environment, supporting technical infrastructures, and most importantly supporting social and organizational environments within the city or institution.

Given the rapidly changing and dynamic aspects of IoT systems and the increasingly complex and resource-constrained environments in which they are deployed – and the number of variables that are outside of an institution's or city's control – the manageability of an IoT system or systems becomes critical for systems success (or failure).

The Potential Benefits of IoT Systems

The potential benefits of appropriately selected, procured, implemented, and managed IoT systems are substantial. Universities and institutions can benefit from IoT systems such as traditional building automation systems (e.g., heating ventilation and air conditioning (HVAC)), energy management and conservation systems, building and space access systems, environmental control systems for large research environments, academic learning systems, and safety systems for students, faculty, staff, and the public. Cities also benefit from IoT systems supporting public safety (e.g., surveillance of high crime areas), air quality monitoring by sector, transportation control systems, city accessibility guidance and support, and many others.[1]

In automotive and transportation systems, IoT can enable health checks of automotive components, Global Positioning System-based location monitoring, route optimization, crash prevention, car-to-car communication, and real-time traffic analysis. City governments and institutions can use traffic data for more effective city planning. In health systems, whether lifestyle, recreational, or patient monitoring for critical functions such as blood pressure, glucose levels, heart rate, or others, IoT devices and supporting systems can monitor, analyze, report health data, and even directly provide appropriately dosed medicine to patients.

Sensor-based analysis in retail spaces can provide business owners valuable analysis of customer behavior and buying patterns, reducing waste, and enhancing profitability.[2] Institutions and cities can use arrays of IoT devices, sensors, and actuators to monitor and analyze buildings, campuses, and spaces for energy usage. With this data, opportunities for increased energy efficiency can be identified. Regulatory and compliance requirements such as carbon emissions requirements can be measured, reported, and enforced. Further, aspirational objectives around carbon emissions, other air quality measures, water contaminant levels, and others can be recorded, studied, and reported.

Systems Loss

> Oh, ye seekers after perpetual motion, how many vain chimeras have you pursued? Go and take your place with the alchemists.
>
> **Leonardo Da Vinci**[3]

As with systems in nature, in social/societal organizations, and particularly in sociotechnical organizations – of which most modern societies are – there is always systems loss. Sociotechnical systems can have many components, facets, and attributes – there can be plans, intentions, resources, alignment, conflict, lag, cause and effect, uncertainty, and surprises. One thing is certain though – there is *always* system loss. Nothing is free. In the excitement, novelty, complexity, and hype around IoT and IoT systems, often what is not realized by cities and institutions is that there is still systems loss. Worse, not only is there systems loss, this loss is substantial and will directly and negatively impact the opportunity for successful implementation of the IoT system.

Systems Loss in Nature

The formalization of the study of systems loss has a rich history of study and publication. It is hard to imagine that the advances in science, medicine, technology, and society that we witness today would have been possible without the discovery, formalization, and documentation of systems loss.

From a scientific viewpoint, the quintessential study of the development and application of systems loss can be found in thermodynamics and, particularly, the second law of thermodynamics.[4]

In the early 19th century, French military engineer, Sadi Carnot, built on some of the work of his father, Lazare Carnot, and introduced the idea of an idealized heat engine. (Among other things, Lazare Carnot is also known for appointing Napoleon as the general-in-chief of the Army of Italy, subsequently being named Minister of War by Napoleon and later as Minister of the Interior by Napoleon).[5]

In his book, *Reflections on the Motive Power of Fire,*[6] (Sadi) Carnot abstracted out the core components of steam engines of the day into an idealized system so that consistent math could be performed in the context of these "heat engines (Figure 1.1)."

In the course of this, Carnot introduced the idea of a heat transfer pattern cycle, subsequently named the Carnot cycle.[7] In this abstraction, he showed that *there is always system loss* – that even when disregarding the effects of friction and machine imperfections (which also cause loss), Carnot proved that there is a maximum efficiency, well less than 100%, of *any* engine. That is, regardless of the machine (engine) or the type of fluid on which the engine runs – whether steam, gas, or others – *a portion of the energy added to the engine will not be converted to work.* That is, *a portion of the energy added will always be lost.* (This is also consistent with a concept that Leonardo Da Vinci had introduced that a perpetual motion machine is impossible.)

Rudolf Clausius,[8] a physics professor at the Artillery and Engineering School in Berlin in the mid-19th century, extended upon Carnot's contributions by formulating the second law of thermodynamics. In his 1850 paper, *On the Moving Force of Heat,* he introduced the early concepts of the second law of thermodynamics.

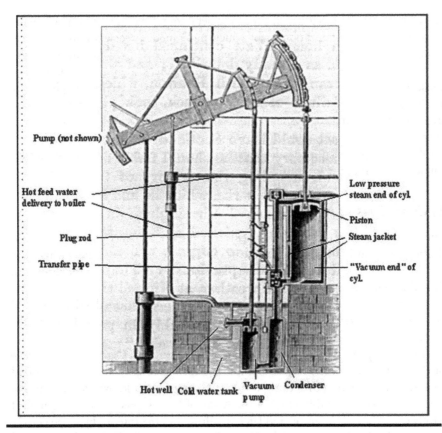

Figure 1.1 A James Watt steam engine, similar to those studied in Sadi Carnot's Reflections on Motive Power of Fire. (From Thurston, Robert H. English: A Schematic of Watt's Steam Engine Printed in a 1878 Book. 1878. Thurston, Robert H. History of the Growth of the Steam Engine. D. Appleton & Co. 1878. https:// commons.wikimedia.org/wiki/File:Watt_steam_pumping_engine.JPG.)[11,12]

In 1865, Clausius gave this irreversible heat loss a name – entropy. The broad concepts are that, left on their own, in systems involving heat (and all do), everything gets cooler, and more generally in all systems, everything tends towards disorder. There is always loss in the system – not everything that goes into the system produces useful output or work.

Systems Loss in Societal Systems – Warfare

An example of systems loss within complex social systems is that of warfare. Because of its complexity, lessons learned in millennia of warfighting can offer some clues to planning for, implementing, and managing complex sociotechnical systems in complex societal groups such as cities and institutions.

> Everything in war is simple, but the simplest thing is difficult. The difficulties accumulate and end by producing a kind of friction that is inconceivable unless one has experienced war.
>
> **Carl Von Clausewitz**, *On War*

Carl von Clausewitz, the famous Prussian war theorist of the 19th century, introduced the concept of friction in war. He describes friction as "the force that makes the apparently easy so difficult."

This notion of friction being things that are "apparently easy" but that are actually difficult in reality can apply to large IoT systems implemented in institutions and cities. At face value, deploying one sensor in a location, routing data over a network, aggregating that data for subsequent processing, analysis, distribution, and consumption should be straightforward and easy. However, doing that 100, 1,000, or 10,000 times on a network (or networks) that is not has homogenous and predictable as originally thought, with resources less than originally anticipated, with vendor support that is not quite what was promised when the deal closed, and surprises, both small and large, will start to reveal cracks. These cracks, in turn, aggregate to "friction," with its detrimental effects on system implementation and management success. Continuing some of the metaphors established in the study of warfare,

> Friction may be self-induced, caused by such factors as lack of a clearly defined goal, lack of coordination, unclear or complicated plans, complex task organizations or command relationships, or complicated technologies.[9]

There are strong metaphors and parallels from which to learn in the study of warfare to implementing these complex technology systems in our complex societal organizations such as cities and institutions.

Another relevant metaphor to draw upon, per Clausewitz and modern military doctrine, is that in warfighting it is desirable to reduce friction, but there is no expectation that it will be eliminated. In fact, *effective military organizations plan and train to operate in the presence of friction, knowing that it cannot be entirely eliminated.*

Systems Loss in Management Systems

Another example appears in management theory. In introducing the idea of systems thinking in analyzing business operations, tactics, and strategy, Peter Senge says in *The Fifth Discipline*, "The irony is that, today, *the primary threats to our survival, both of our organizations, and of our societies, come not from sudden events but from slow, gradual processes.*"[10] This is yet another example of nondescript systems loss that quietly aggregates and ultimately directly impacts the chances of IoT systems implementation success in terms of both ROI and cybersecurity posture.

Senge goes on to describe a "learning horizon" and our ability to observe and develop internal models of cause and effect.

"But what happens when we can no longer observe the consequences of our actions? What happens if the primary consequences of our actions are in the distant future or are in a distant part of the larger organization in which we operate? ... When our actions have consequences beyond our learning horizon, it becomes impossible to learn from direct experience."[10] IoT systems also stretch across these large organizations – institutions and cities – and the comprehensive impact is never felt in any one place.

"We learn best from experience but we never directly experience the consequences of many of our most important decisions."[10] Similarly, because the full impact of IoT systems across cities and institutions is not comprehensively felt in any one place, learning from any particular IoT systems implementation is challenging.

This has strong parallels to where we are as institutions and cities. To add value and be competitive with other institutions and cities, we are rapidly buying and deploying IoT systems of increasing scale, and yet we really have limited basis for making good decisions about them. We're pressured into making decisions about deploying our nth IoT system, but we have little feedback from our early IoT systems deployments that can help influence that decision.

Systems Loss in IoT Systems

In IoT systems, loss occurs at the device level in IoT systems through unexpected installation costs, failures, misconfiguration costs, and others. Loss occurs at the network and network segment level through challenges of device enumeration and management, miscommunications between network managers (e.g., the central IT organization) and the IoT system owner around network segmentation and management, and others. Other losses occur in the organizational coordination level. Loss also occurs in resourcing. At each individual point or region, the loss may be small and not obvious. However, those losses aggregate and become substantial. Because of the rapid proliferation of IoT devices, the network segments supporting them, and the complexity of coordination of developing and maintaining organizational support resources, that system loss becomes significantly larger.

This systemic loss is insidious. It doesn't show up and knock on one's door advertising itself as loss or as a problem. Rather, it quietly, and initially imperceptibly, aggregates until system performance is substantially degraded, resourcing estimates prove themselves to be inadequate, uncertainty is significantly increased, cyber risk increases – perhaps greatly, and system manageability is degraded to the point of lost utility or failure. The scale and rate of growth of IoT devices, network segments, and systems greatly exacerbate this loss.

These analogies in other complex societal endeavors, such as warfighting, natural physical systems, and management organizations, can act as strong reminders and possibly even, in part, provide the basis for building partial models to help us frame the implementation of complex sociotechnical systems such as IoT in a substrate of complex societal organizations.

IoT Systems Manageability

Because systems loss is inevitable – not the least of which in complex IoT systems in complex societal organizations – developing approaches and methods to manage this loss, and most importantly, to be aware of this loss and learning to operate with that loss are critical to the success of IoT systems in cities and institutions. In turn, this is critical to the success of the smart cities or smart campus concept.

As leaders and administrators in institutions and cities, we can tend to get lost in the glitter and bling of potential promises of technology and not fully grasp the challenges of administering these complex systems.

Ironically, this is also true of some technologies that are offered to ostensibly mitigate IoT systems risk. The reason that these ostensible risk mitigation technologies are often ineffective is that they tend to vastly oversimplify the problem of IoT systems risk mitigation. These risk mitigation technologies often present themselves in neat technology packages and appear very convenient. While there are definitely technology components that can help with IoT systems risk mitigation, proposals or sales pitches that a single technology solution can address all IoT risk mitigation issues can be the equivalent of selling snake oil.

However, there are things that institutions and cities can do to mitigate this systems loss and increase the likelihood of successful IoT system implementation and operation. While there are many aspects to choosing, implementing, and operating IoT systems, a recurring thread and theme is that of *systems manageability*. By being aware of and demanding high degrees of systems manageability in the IoT systems that institutions and cities acquire, deploy, and operate, the opportunity for positive ROI's and non-degraded institutional cyber risk profiles are possible.

> Heeding and developing institutional maturity for identifying and demanding highly manageable IoT systems is critical for success.

It is one of the most important factors that cities and institutions can control in this rapidly increasing complexity of our societal and institutional environments, the exponential growth in the number of IoT devices, the systems that support and integrate them – all combined with the limited human, technical, and fiscal resources that almost all institutions and cities face that overwhelm our traditional approaches to enumerability, inventory, classification and categorization, and generally risk mitigation and management.

Systems manageability is perhaps the primary factor in addressing the slow bleed – that systems loss, the entropy – of the complex sociotechnical environments in which we live. Our societal/social systems are already complicated, but the integration and embeddedness of technology only expand that complexity. The manageability of an IoT system has direct and substantial implications on these limited human, technical, and fiscal resources.

There are many components and aspects to IoT systems manageability. The first step in discerning an IoT system's manageability is discerning the system's "knowability." What do we know about these systems that we are deploying in and around our institutions and cities? Do we just open the doors to our physical spaces and networks and let a third party install whatever, wherever? Or do we seek to know what is entering our physical and network environments? Discerning a system's knowability is a prerequisite to determining a system's manageability.

Some examples of these systems' knowability components and attributes include:

- Does the system have a name shared by all stakeholders?
- Who are the stakeholders in the consuming institution or city?
- Is there a primary point of contact within the institution/city for that system?
  - A coordinator at the vendor side?
  - A coordinator at the city/institution side?
- What are the expectations of data produced by the IoT system by *all of the stakeholders – more likely than not that there will be different expectations of data?*
- How well known is the system?
  - New vendor? Known vendor?
  - Degree of trust with vendor?
  - Documentation?
  - User and system support training?
  - How many endpoints, e.g., sensors and actuators, are there in the system?
    - Where are they? Do we know the location?
    - What is the IP address? The MAC address?
    - What is the current firmware version?
    - How are the devices updated/patched?
    - Are they updated/patched?
  - What are indications of health of these endpoint devices?
    - How does a healthy device present itself?
  - What is the central managing/controlling/aggregating application of these devices?
    - Does it sit on an on-premise server? Software-as-a-Service (SaaS)? Other?
    - What are the requirements of this application and supporting server?
    - How many servers?

- What are the desktop client applications involved with this application?
  - What are the requirements?
  - How many are there?
- Is there a risk agreement between the provider and the institutional consumer?
  - What is the mutual ownership between system success and failure?
  - Is there mutual ownership? Or is the client on their own?
- Others.

This is just a subset of IoT systems *knowability*. How much can we know about a complex IoT system in our complex institutional and city sociotechnical environments?

Capturing this information (and other information not enumerated here) is involved and resource-intensive.

We'd like to capture as much as we can within the bounds of our limited resources. But, importantly, if – when – we can't capture it all, *we want to know – and admit to ourselves – that we haven't captured it all and that capturing it all may indeed be impossible.* We acknowledge that we will be working with incomplete information and plan for this and work with this.

The institutional internal view and reflection that – though we seek to capture as much information as possible but acknowledge that we will always be working with partial and/or incomplete information – is a much better approach than extending the fantasy that we have complete knowledge of our systems. Making the assumption that we can or have captured, counted, and enumerated all aspects of the IoT system is a critically flawed basis for developing management and mitigation strategies for the institution or city.

We have to admit to ourselves as institution and city leaders that it is probable that many of *our sociotechnical IoT systems have become feasibly non-enumerable.* That is, we really don't know what's there. And, further, if we keep building management and risk mitigation approaches on this assumption that we can count everything, that we know where everything is, and we know what everything does – then we've got problems. Again, we seek to know as much as we can about our systems, but we must come to terms with the fact that we will not capture everything.

Given this tough state of affairs regarding just knowing the systems that we are deploying in our cities, institutions, and corporate campuses, discerning systems manageability – which has systems knowability as a requisite – is even harder.

Towards IoT systems *manageability*, within our institutions and cities – we want to use knowledge about systems, knowability, as a strong basis and then we want to look at internal resources available, skill sets available, market availability of skill sets, contract agreements, projected and actual vendor support, communication and coordination between and within institutional organizations and departments and others. This is no small task. Some components of IoT systems manageability include:

■ Do the stakeholders, and their respective staffs, in city and institutional IoT system(s) have capacity to participate in the oversight, management, risk mitigation of these systems?
  - What current tasks and roles will they give up to participate in this work?
  - What skill sets do the staffs of the stakeholders need to support the stakeholders?
  - What skill sets are available in the institution or city to interpret the data so that it is actionable and usable in the context in which it is produced and consumed?
■ What staff time is available to manage the performance of IoT systems vendor contracts?
  - What will these staff need to give up to manage these vendor system contracts?
■ For the hundreds, thousands, or more of endpoints – devices deployed, who is going to support those devices?
  - What organization or department will support those devices?
  - Does that organization or department know that they are tasked with supporting these devices?
  - What is the existing budget and effort capacity to do this work?
    • Is it planned or is this a surprise?
    • What other work does the supporting organization give up to do this work?
    • Where is the resourcing – staff, funding, and scheduling – coming from to get this work done?
■ Regarding systems applications
  - Who supports the server applications or SaaS applications?
  - Who establishes, adjusts, and maintains the configurability?
  - Who supports the client (e.g., desktop) applications?
  - Who trains the users?

Identifying *IoT systems manageability* as one of the top priorities, if not the top priority, for IoT systems selection, procurement, implementation, management, and even systems retirement – is essential. It is critical that systems-consuming institutions and cities see this and that *they create demand from IoT systems providers for more manageable IoT systems* and importantly that the provider/vendor shares the effectiveness of that objective with the consumer.

Ecosystem and Market – IoT Systems Consumers and Providers

IoT systems institutional and city consumers and their provider partners make up an ecosystem of IoT systems. While they are distinct entities in one sense, much of their operational lines are increasingly blurred between institutional/city

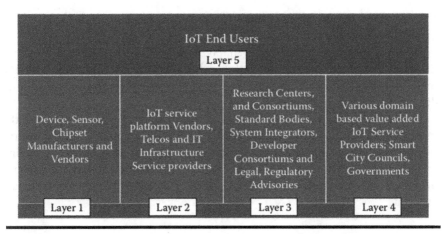

Figure 1.2 Complex IoT ecosystems. (Courtesy of Abhik Chaudhuri, Internet of Things, for Things, and by Things.)

consumers and their technology systems providers to include network boundaries, support agreements and roles, risk sharing, performance contracts, and others (Figure 1.2).

Because of this, IoT systems vendors must also be motivated by developing not just IoT systems, but also IoT systems services and support mechanisms. Selecting, deploying, and managing IoT systems are still nascent for most cities and institutions. Mistakes will be made. One of the key places where mistakes are being made is estimating and planning for the resources and skill sets required to deploy and operate these systems. While this is problematic, over time, these institutional consumers will learn from these mistakes, mature in their IoT systems acquisition approach, and begin demanding better systems. In particular, the demand for better IoT systems will drive the demand for better IoT systems manageability.

Organizations seek to be aware of, manage, and reduce this systems loss by selecting for more manageable systems – in both the technical and the sociotechnical senses – and will see much greater value add of their IoT systems than those of their peers or competitors.

Similarly, those IoT systems vendors and providers acknowledge this guaranteed systems loss within a consumer city or organization and help them reduce that loss through increased systems manageability – while acknowledging that there will always be some loss – *will outperform and outlive their competitors.*

Because of the unbridled potential of new IoT systems coupled with our lack of experience as city and institutional consumers for establishing performance expectations, vetting criteria, and IoT systems deployment and operational experience, there are many, many IoT systems on the market that have very limited value and, in fact, can have negative value by causing lost investment and degrading a city's or institution's cybersecurity posture. In a nutshell, there is a lot of IoT systems garbage on the market.

In time, some of these ill-conceived, poorly defined, and poorly supported systems will shake out as being not useful and possibly harmful. But this will take some years, and along the way, damage will be done in terms of lost ROI and degraded cyber risk profiles.

Those IoT systems providers that can grasp the complexity of matching the institutional consumer need for the system, along with comprehending the institutional consumer's complex technical and social/societal substrate (often including working with and within a large bureaucracy) – into/onto which that IoT system is deployed – will have a substantial market advantage and will bring greater impact over time.

Approach

This book will focus on the relationship between this technology and that of major societal organizations such as cities and universities. These major societal organizations have a duty and obligation to serve, protect, and enhance the lives of the people that live and work within them. As such, one of their obligations is to identify and seek to manage and mitigate risk to their constituencies and organizational structures.

The book is composed of chapters that cover various aspects of IoT systems and risk mitigation and cybersecurity around the same. Within each chapter, a particular aspect or phenomena of IoT systems in cities and institutions will be discussed. The intent is to provide some language and conceptual frameworks for the issue. Our shared language about these IoT systems must evolve and do so quickly, if we are to successfully manage these systems and manage and mitigate risk around the same. Some chapters will also include proposed mitigation steps and actions. None of these approaches are written in stone, and there are many ways to accomplish the objectives, but these will be worthy of consideration for your own city or institution. Similarly, commercial providers of IoT systems products and services can create competitive advantage by helping institutions and cities solve these complex challenges.

Chapter 2, "Differences between IoT and Traditional IT Systems," discusses how IoT systems are different from traditional enterprise IT systems within institutions and cities.

Chapter 3, "Defining IoT Systems Implementation Success," provides criteria for analyzing the success (or not) of IoT systems implementations.

Chapter 4, "Systems of Systems and Sociotechnical Systems," introduces IoT systems as both systems of systems and sociotechnical systems.

Chapter 5, "Systems Seams, Boundaries, and the IoT Ecosystem," discusses systems losses at organizational boundaries and how those losses aggregate in resource-constrained environments, such as most institutions and cities, to directly impact the likelihood of IoT systems implementation success.

Chapter 6, "IoT Systems Manageability," describes systems manageability in more detail, and the positive impact of strong systems manageability can have mitigating systems losses in resource-constrained environments in institutions and cities.

Chapter 7, "IoT Systems Vendor Relations & Vendor Management," covers the critical relationship of the city or institution and the IoT systems vendor or provider.

Chapter 8, "Templates for Institutional & City IoT Systems Planning & Operations," offers some templates for planning and implementing IoT systems.

Chapter 9, "Strategy Implementation," presents a high-level strategy for selecting, deploying, and managing IoT systems within the institution or city.

Suggested Reading

1. Cleveland, Robin, and Carolyn Bartholomew. Hearing on China, the United States, and next generation connectivity. U.S.-CHINA ECONOMIC AND SECURITY REVIEW COMMISSION, 2018, 132.
2. Chaudhuri, Abhik. *Internet of Things, for Things, and by Things.* Boca Raton, FL: CRC Press/Taylor & Francis Group, 2019.
3. sataksig. Perpetual Motion Machines and the Search of Free Energy. Earth Buddies (blog), November 26, 2017. https://earthbuddies.net/perpetual-motion-machines-and-the-search-of-free-energy/.
4. Second Law of Thermodynamics. In *Wikipedia*, December 25, 2018. https://en.wikipedia.org/w/index.php?title=Second_law_of_thermodynamics&oldid=875356570.
5. Carnot, Lazare. In *Wikipedia*, November 20, 2018. https://en.wikipedia.org/w/index.php?title=Lazare_Carnot&oldid=869870374.
6. Carnot, Sadi. "Heat Engine" *Reflections on the Motive Power of Fire: And Other Papers on the Second Law of Thermodynamics*, by É. Clapeyron and R. Clausius. Mineola, NY: Dover Publ., 2009.
7. Carnot Cycle. Accessed January 3, 2019. http://hyperphysics.phy-astr.gsu.edu/hbase/thermo/carnot.html; https://en.wikipedia.org/wiki/Carnot_cycle.
8. Clausius, Rudolf. In *Wikipedia*, November 13, 2018. https://en.wikipedia.org/w/index.php?title=Rudolf_Clausius&oldid=868579650.
9. Warfighting, Mcdp1.Pdf. Accessed January 3, 2019. https://clausewitz.com/readings/mcdp1.pdf.
10. Senge, Peter M. *The Fifth Discipline: The Art and Practice of the Learning Organization.* Rev. and updated. New York: Doubleday/Currency, 2006.
11. Thurston, Robert H. *English: A Schematic of Watt's Steam Engine Printed in a 1878 Book.* 1878. https://commons.wikimedia.org/wiki/File:Watt_steam_pumping_engine.JPG
12. Thurston, Robert H. *History of the Growth of the Steam Engine.* D. Appleton & Co. 1878. https://commons.wikimedia.org/wiki/File:Watt_steam_pumping_engine.JPG.

Chapter 2

Differences between IoT and Traditional IT Systems

What Is an IoT System?

Internet of Things, or IoT, systems are much of the underpinnings of smart cities and smart campuses. These systems have the potential to greatly impact society and offer substantial benefit to institutions, cities, governments, and the populace in general. However, if these systems are not selected, procured, implemented, and managed thoughtfully, not only will that potential benefit not be seen, but these systems can substantially increase the cyber risk that an institution or city faces. Poorly implemented and managed, they can also be an opportunity for substantial financial and reputational loss and even injury and loss of life.

IoT systems are complex sociotechnical systems. The systems include thousands or millions of networked computing devices along with more centralized servers, to include cloud-based services, for data aggregation and sometimes command and control, embedded in the spaces around us. There can be a complex interplay, whether immediately or with temporal delay, between these systems and the humans that work, study, and play in their midst.

These IoT systems do not exist independently – they require a substrate onto which they are deployed to support them. This substrate is comprised of the existing and often evolving network architectures and technical infrastructures of cities and institutions. These network systems are, themselves, complex systems with multiple parts, components, architectures, and management requirements. Couple that with the fact that the large, bureaucratic, and multifaceted organizations and departments that occupy the same space as deployed IoT systems are also complex

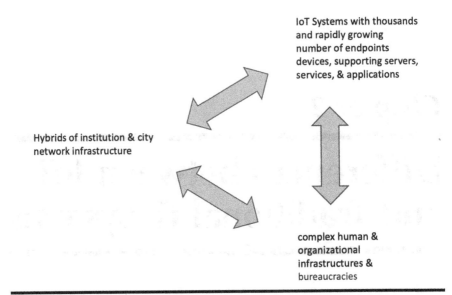

IoT Systems with thousands and rapidly growing number of endpoints devices, supporting servers, services, & applications

Hybrids of institution & city network infrastructure

complex human & organizational infrastructures & bureaucracies

Figure 2.1 Mutually interacting complex systems.

organizational systems with all of the complexities of human and organizational behavior in the mix (Figure 2.1).

This creates an, often not well defined or explained, interdependence of complex systems.

- The complex IoT system with its thousands or millions of nodes and their associated behaviors and functions interacts with …
- The complex substrate, often simultaneously existing/legacy and evolving, upon which the IoT needs to exist interacts with …
- The complex institutional and population structures of bureaucracies, cities, and institutions interacts with …
- Back to the top—the complex IoT system ….

Six Core Differences

IoT systems and traditional enterprise IT systems for the institution or city differ in at least six ways:

1. Raw magnitude of networked computing devices and rate of growth of the same
2. Variability in types of devices *and* variation in hardware and software components comprising each device, where each device is a "Thing" in Internet of Things

3. Lack of shared language for describing IoT systems; lack of categorization or classification of the same
4. Organizational spanning characteristics of IoT systems
5. Tendency for IoT devices and systems to be "out of sight, out of mind"
6. Lack of precedent for the implementation of these systems in cities and institutions.

Big Numbers

A white paper from Cisco in 2011 suggested that there would be 50 billion networked computing devices by 2020.[1] Since then, that number has fluctuated, but the general consensus remains in the low tens of billions of devices. Further, the rate of growth in the count of those devices appears to be exponential.

For an institution, such as a university or a city then, it follows that they are experiencing and will continue to experience very large numbers in their domain. Similarly, the institution or city will likely experience a very rapid, probably exponential for a few years, rate of growth.

The issue is that does the institution have the inherent management capacity to

1. Handle the *raw magnitude of number of devices* at any point in time
2. The *rate of growth in the number of devices*—does the institution have the capacity for rapid change in management ability?

The answer is almost certainly "No" for any real-world, non-idealized, institution or city.

Higher Education institutions, for example, are working hard (and with increasingly reduced funding) to address old, existing problems—modernizing legacy systems (and addressing cultural change for the same), responding to new cyber threats and vulnerabilities, adapting to new cyber threats and vulnerabilities, responding to organizational changes brought on by new technology implementation, and others. There is not a coincidental, mystical increase in systems management capacity that comes along with this systems and environment change. The rapid growth in the number of IoT devices, their corresponding IoT systems, and the growing interdependencies between them increases the systems management requirements of institutions and cities, most of which have very real resource constraints for managing these and other systems.

High Variability and High Variation

IoT systems and devices vary in both the variability of deployed IoT devices – number of types of devices – and the variation in the hardware and software components that comprise all of those endpoint devices.

The ongoing miniaturization of electronics, the incredible drop in cost of many of these electronics components, the availability of the same stemming from globalized trade, and other factors drive incredible innovation in the space of IoT devices and systems. A small sample of examples include:

■ FitBit fitness "watches"
■ Building automation systems
■ Video surveillance systems
■ Glucose pumps
■ Pacemakers
■ Energy management systems
■ Livestock monitoring, e.g., the "MooMonitor+"[2]
■ Precision farming
■ Drones, drone management, drone defense[3]
■ Wi-Fi-connected wine bottles[4]
■ Home security
■ Many others

The wide variety of devices and systems and rate of innovation of new systems makes it very difficult to categorize or classify these devices and systems.

Variability of Devices

Categorization and classification are staples of traditional risk management approaches. Historically, we create or identify groups of similar things and seek to manage them as buckets of similar risks – similar likelihoods of events and similar impacts of events. (The negative events are usually the ones that we're most interested in dealing with.)

A challenge of trying to analyze and manage risk with this traditional approach is: what are the buckets? How do we group these things? What things are similar and what are different? Cities will have a vast array of these types of systems and devices. Institutions such as universities and colleges will likely have even more with their diverse activities and service lines. National Collegiate Athletic Association (NCAA) IoT-based football helmet protection technology[5] might exist within a city block of IoT-based advances in clinical healthcare which might exist near substantial research facilities and all of which working under the auspices of Building Automation Systems (Figure 2.2).

Variation within Devices

As discussed, the variety of types of devices and of the hardware and software components within each device is very high. IoT devices do numerous different

Figure 2.2 University of Washington, just one example of very different types of IoT systems working next to each other.

tasks, including measuring building energy, video monitoring a space, reading a heart rate, and sensing air quality every few seconds in a research facility.

A different type of difference is the *variation* in the types of components within each device. Devices can have many different types of hardware from many different manufacturers as well as many different layers of software, each possibly from a different software company (or person). This huge variation contributes to the challenge of identifying device categories that can be helpful in developing risk management approaches. This variation of device components across all of the deployed devices also makes provenance tracking and provenance management very difficult. That is, it can be very difficult to discern and know what the source, manufacturer, or developer is of each of the components within each of the deployed IoT devices. It can be, and usually is, difficult to know what companies or people participated in the manufacture all of the hardware and software components within the many different deployed IoT devices (Figure 2.3).

In their paper, *Internet of Things Device Security and Supply Chain Management*,[6] researchers Lee and Beyer contribute:

> "... policies relating to electronic supply chain security at national level are lacking ... although companies try their best to follow piece-meal governmental and industry guidelines for supply chain security, this vigilance is only as strong as a company's dedication to security."
> [Supply chain policy shortcomings] "... arise because cybersecurity issues are highly complex and difficult for policymakers and industry leaders to reach agreement upon."

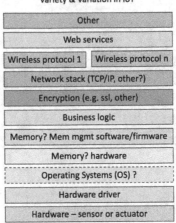

Figure 2.3 **Variation of components within an IoT device.**

Lack of Language

Related to the high variability of types of devices and systems, there is also a current lack of shared language, mental constructs, and frameworks to describe, discuss, and analyze these systems across individuals and organizations within an institution or city. Without the shared language or experience to discuss these, establishing a mechanism to make good decisions about these systems—to include mitigating risk stemming from their deployments—is virtually impossible.

Organizational Spanning Nature of IoT Systems

IoT systems almost always span multiple organizations within an institution or city. For example, an IoT system such as a complex environmental control system in a new university research laboratory will involve many institutional organizations—both within the academy and within the administrative component of the institution. The end-user research laboratory Principal Investigators (PIs), the research support staff, and graduate and undergraduate students will certainly be involved as they will likely be direct users of the system. Also involved will be a large array of university administrative support to include:

- A capital development group might be involved to build the laboratory with the embedded, research-supporting IoT system
- A facilities management group will be involved to support the lab – to include its traditional HVAC needs, IoT system needs, and other building management needs

- The institution's central IT, providing the network backbone, possibly network segmentation schemes, and others
- Local IT support that perhaps was embedded with the academic research group or other local IT support
- Many vendors, contractors, subcontractors, and third-party suppliers
- Others.

All of these separate organizations within the institution create gaps and spaces for questions of systems oversight, accountability, ownership, and expectations of performance to fall. It is typically not clear who owns the whole system or who owns what components of the system. Often, individual organizations assume other organizations have accountability for the system when they actually do not. For example, the PI researcher may assume that "central IT is making sure that this system is deployed properly and is secured to best practices," while at the same time, the central IT group may have no idea that the system is even being installed. Cities have very similar organizational complexities.

Out of Sight, Out of Mind

Because it's so new and so different, it is often forgotten that the Thing in IoT is a device that computes and is networked. The T in IoT is indeed a reasonably powerful networked computer. However, because these Things are typically embedded in the environment around us—and not in the traditional sense of an office workstation, laptop, or a server in a data center—these almost countless networked computers – IoT devices – are not thought of in terms of needs of traditional computer support or risk mitigation, e.g.,, an electrical power meter mounted in a utility closet or boiler room that samples electrical load every 5 min that looks nothing like a traditional computer (Figures 2.4 and 2.5).

Lack of Precedence for Implementation

Institutions and cities generally have limited to no experience with IoT systems implementations. Because the IoT phenomenon is nascent, yet evolving so rapidly, there is little in the way of institutional knowledge of previous system installations. Further, while it's beneficial to share lessons learned, at this point in time, other cities and institutions also do not have depth in experience in implementations either. They don't have years of post-implementation experience to analyze these systems and thus do not have institutional knowledge about these systems developed. Similarly, competitors also don't have years of IoT systems deployment and operations experience from which to draw through observation.

There have, of course, been installations of HVAC systems, elevators, and similar that use some kinds of remote automation and control. However, the scale of the networked computing devices of the IoT phenomena is orders of magnitude

Figure 2.4   Energy measurement the old way.

Figure 2.5   A networked industrial energy meter in an institutional building.

higher than these historical systems. Also, most of these historical systems have typically not been networked (e.g., IP/TCP controlled)—certainly not systems over 25 to 30 years old.

Deeper Dive into Variability and Variation

The aspects of both *variability* of the number of different types of IoT devices (and the rapid increase of the same) **and** the large *variation* of the hardware, firmware, and software component types between each of those deployed devices within the institution or city warrant a deeper dive.

There is a large variety of types of IoT devices and systems. We've discussed some already:

- Energy management systems and devices
- Fitness devices
- Entertainment systems and devices
- Clinical systems and devices
- Audio/video systems and devices
- Security/safety systems and devices
- Research systems and devices
- Teaching and learning systems and devices
- Industrial control systems and devices
- Building Automation Systems (e.g., HVAC)
- Toys
- Many others.

This is quite a spectrum, and the spectrum will only continue to broaden and populate. This wide variety of types of IoT devices and systems does not lend itself well to clumping different systems/devices together into buckets for traditional risk analysis.

Risk Buckets

The traditional mechanism for managing risk is to group like items into subsets and then develop estimates of likelihood of event and impacts of those events. In financial risk management, risk buckets could include similar financial instruments. For example, a bond portfolio might have buckets of 5-, 10-, and 30-year maturities.[7] Similarly, another portfolio might have a bucket for value stocks and another bucket for growth stocks. The point is that similar bonds or stocks are placed into the same bucket across a portfolio to establish a perspective or spectrum of risk across the portfolio. Without such a bucket mechanism, we seem to be left with the extremes of (1) trying to evaluate every single stock or bond in the portfolio, or (2) dumping everything into one giant bucket – which isn't a lot of help (Figure 2.6).

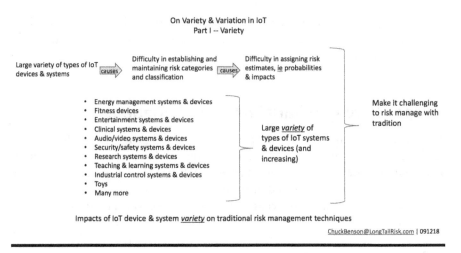

On Variety & Variation in IoT
Part I -- Variety

Large variety of types of IoT devices & systems → causes → Difficulty in establishing and maintaining risk categories and classification → causes → Difficulty in assigning risk estimates, ie probabilities & impacts

- Energy management systems & devices
- Fitness devices
- Entertainment systems & devices
- Clinical systems & devices
- Audio/video systems & devices
- Security/safety systems & devices
- Research systems & devices
- Teaching & learning systems & devices
- Industrial control systems & devices
- Toys
- Many more

Large *variety* of types of IoT systems & devices (and increasing)

Make it challenging to risk manage with tradition

Impacts of IoT device & system *variety* on traditional risk management techniques

ChuckBenson@LongTailRisk.com | 091218

Figure 2.6   Large variety of types of IoT devices and systems.

The challenge with this risk bucket approach with IoT systems and devices is that it's not obvious how to determine the buckets, nor is there a broadly shared approach on what those risk buckets are. A larger agreed-upon basis of examples across a population, i.e., something that lends itself to developing actuarial tables or other analyses, is not available for IoT systems and devices.

Law of Large Numbers

The law of large numbers[8] tends to guide us to more accurate predictive results across large numbers of things or events. At this point, we generally don't have these large numbers of things or events for IoT systems implementations available in an actionable sense. While there is plenty of data, and more coming, being generated by IoT *devices*, we don't have a lot on histories of IoT *systems implementations*. Because of this, there is limited predictive value for this approach.

This problem of lack of large numbers data for systems implementations will change as we have more and more implementations of IoT systems. However, the variability problem will still plague us. We'll have lots of instances of device deployments, but they'll be spread across this variety spectrum and analyzing risk in the traditional sense will still be challenging.

Back to Risk Buckets

So, the challenge before us is either to find ways to group these nuanced, sometimes weird, and rapidly evolving systems and devices, or to find a new way to manage risk. The approach to the latter – finding a new or adapted way of managing risk for IoT systems deployed in cities and institutions – is fairly new territory and not particularly easy, though it is important that we not rule it out. We should think broadly about alternative approaches and test them. And different from the early days of Edward Lloyd where his coffee house was used to gather rumor, news, and intelligence on its/his path to becoming Lloyd's of London,[9] we do have the ability to simulate and test many different alternative approaches.

In the immediate and near future, however and until we develop a different approach, we will probably have to stick with the old-school method of risk buckets in order to make an initial foray into risk mitigation for IoT systems.

So, then, how these IoT systems and devices group in this traditional method of establishing groupings of different types of devices and systems into different risk buckets – each with their own characteristic likelihoods of events and impacts of events? Do we group these:

- By device manufacturer?
- By function?
  - What is "the function?"
  - What are the expectations?
- By physical size?
- By network bandwidth?
- By IoT system provider?
- By deployment geography?
- By installation difficulty?
- Many, many more possibilities ….

There are an almost limitless number of possibilities of criteria for creating risk buckets. That number of possibilities becomes yet substantially bigger when combinations of criteria are considered. The question is figuring out how to use this traditional approach in establishing a framework for mitigating and managing IoT risk.

> Many forms of Government have been tried, and will be tried in this world of sin and woe. No one pretends that democracy is perfect or all-wise. Indeed it has been said that democracy is the worst form of Government except for all those other forms that have been tried from time to time....
>
> **Winston Churchill**[10]

I feel that this sentiment is similar to what we face with IoT system risk mitigation of using the risk bucket approach of grouped items of ostensible likelihoods of events and ostensible impacts of events. Because of the many potential bucket types that do not group together in clear or natural ways, the approach can feel fairly arbitrary. However, it's all that we currently have that is broadly acceptable and implementable.

Given these challenges of nonobvious risk grouping options, one approach is to create buckets based on how IoT systems devices are *procured* within an institution or city (Figure 2.7).[11]

IoT devices and systems have multiple entry points into an institution or city. In the case of a university, e.g., any student, faculty, or staff member can simply bring IoT devices such as fitness trackers with them when they go to school or work. They can even bring in small IoT systems, such as Nest video surveillance systems.[12] I call these types "walk-on" devices and systems. They are characterized by not involving any institutional or city/government funding or spending. These walk-on devices and systems just show up on campuses and/or within city boundaries. This could possibly be one broad bucket for IoT system and device risk mitigation.

Another broad bucket, with potential sub-buckets, can be defined as IoT systems and devices that enter the campus or city boundary or jurisdiction via institutional spending of some sort. This means that, in one form or another, institutional or city funds were involved. This bucket can then be broken down into smaller buckets for analysis. For example, a university might have a bucket for IoT systems and devices purchased through their central procurement mechanism. Another of these sub-buckets could be institutional or government spending on IoT systems and devices through an institutional credit card. In this case, institutional funds are still used, but there is typically no central oversight of the purchase outside of perhaps daily and monthly purchasing limits. Another

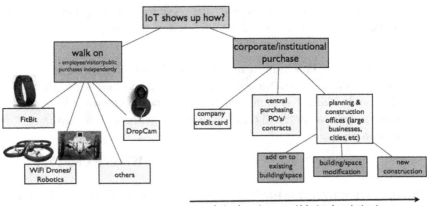

How do IoT systems & devices show up in your networks?

-- A possible method of IoT classification for risk management --

Figure 2.7   One approach to IoT device and system classification.

sub-bucket could be funding for IoT systems and devices that comes as a part of the institution' s or city's major and minor capital development programs. The IoT system purchase might be a part of plan of putting up a new building. Perhaps, it is a new HVAC system that includes traditional heating and air-conditioning as well as carbon dioxide-sensing-based room occupancy detectors[13] coupled with room and building access control systems.[14] This funding could come from capital dollars as a part of constructing the building. Similarly, the remodel of building wing or maybe a laboratory upgrade in a university could involve minor or major capital funds or both.

This procurement-source mechanism for creating risk buckets and categories can serve at least as a starting point for IoT system and device classification.

IoT Systems and Device Taxonomies

Another approach for defining risk buckets for IoT systems is to create IoT systems and device taxonomies for use within the institution or city and ideally an approach to be shared more broadly.

At first glance, creating a taxonomy for IoT devices and systems seems like the natural thing to do. However, creating and using taxonomies for IoT systems and devices can be problematic for a number of reasons:

- Creating IoT system taxonomies is not easy to do
- Diminishing returns of taxonomy development and implementation
- Effective taxonomies can vary between industries, institutions, cities, etc.

The first issue is due to the large variety of types of IoT systems and devices mentioned before as well as a lack of shared language for these systems and devices both within the institution or city and across industries.

The second issue is that it is easy to run down (multiple) rabbit holes in the course of both developing taxonomies and deploying them. Because of this variety in types of IoT systems and devices, developing language and classification and categorizations schemes is hard work. After the hard work, time, and resource investment in developing a classification strategy, even the best schemes have a lack of satisfactory feel and confidence about them. Because this effort is new within this rapidly evolving and landscape changing technology, even well-executed classification development efforts typically have a higher degree of uncertainty about them. This in turn makes deploying the classification scheme or taxonomy challenging because (1) typically a lot of interpretation of IoT systems and device attributes is required and this generates work, and (2) because of the newness and uncertainty of the approach, there is this nagging feeling of "are we doing this right?" which doesn't inspire confidence.

As taxonomies broaden and deepen, their deployments in practice can involve significant time and effort investment and soon hit the point where the potential value of the taxonomy is overshadowed by the work that it took to develop and deploy the taxonomy.

For academic and research use, though, deep complex taxonomies may have great value. The important thing for institutions and cities to remember is that investment in time and resources to development, communicate, and evangelize in their resource-constrained environments can quickly hit a point of diminishing returns (Figure 2.8).

Finally, the third difficulty to the IoT system and device taxonomy development task for an institution is that taxonomies that may be most effective to a given institution or industry may not be same as other institutions or industries. This has two impacts: (1) it makes it hard to learn from another institution or industry regarding their approach, and (2) the opportunity is lost to gather a much broader set of actuarial data.

However, even with these challenges, it is worthwhile to come up with at least a high-level taxonomy or classification and categorization scheme for your institution or city.

Developing these schemes from scratch is particularly challenging for the reasons stated above. Another approach could be to borrow the IoT systems taxonomy of a peer institution or industry. A challenge here is that (1) peer institutions may not have an IoT systems classification scheme to share, and (2) if they do have a classification scheme developed, they may not be willing to share.

Yet another approach to avoid creating an IoT systems taxonomy from scratch is to start with various published reference architectures for IoT systems.

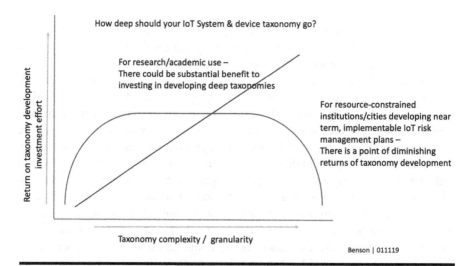

Figure 2.8 Diminishing returns for real-life application of IoT taxonomies.

Some of these are architectures for design, and others also have the specific intent of introducing language and concepts for developing taxonomies. There have been substantial and important efforts to develop reference architectures and taxonomies, and they may, or aspects of them may, work for your organization or city. These architectures can be used as a rich source of language and concepts for developing IoT systems classification, categorization, and systems profiling approaches which then can be the basis for developing risk buckets for IoT systems and devices. Two of these will be discussed here:

- NIST Framework for Cyber-Physical Systems (NIST.SP.1500-201)
- An approach proposed in the proceedings *Internet of Things: a definition & taxonomy.*

The NIST Framework for Cyber-Physical Systems: Volume 1: Overview (NIST. SP.1500-201)[15] is a great document for reference, language, and ideas on addressing IoT/cyber-physical system (CPS) classification and categorization for your context. (Note that the NIST document uses the term Cyber-Physical Systems or CPS. For our purposes, we can consider IoT and CPS to be near-synonymous).
From the document:

> NIST established the CPS Public Working Group (CPS PWG) to bring together a broad range of CPS experts in an open public forum to help define and shape key characteristics of CPS, so as to better manage development and implementation within and across multiple "smart" application domains, including smart manufacturing, transportation, energy, and healthcare … The focus of this Framework is to develop a CPS analysis methodology and a vocabulary that describes it.

In this document, NIST uses the notions of domain, facets, and aspects to help establish the framework. Domains include things like broad industries. Examples are shown in Figure 2.9.

The second notion is that of "facets." Facets, in this context, are views of the IoT system or CPS, and there are three of them: the "conceptualization" facet relates to high-level goals and functional requirements. The "realization" facet relates to detailed design, production, implementation, and operation. And the "assurance" facet provides the feedback loop regarding whether the IoT/CPS actually instantiates what was envisioned in the conceptualization facet.

The "aspects" part of the framework captures broad concerns and issues that may affect multiple other components of the framework. Examples of identified aspects include trustworthiness, data, boundaries, and life cycle (Figure 2.10).

While discussing the derivation of the framework, the document defines the concept of "concern," borrowing from conventions established in ISO/IEC/IEEE 42010,[16] as "an interest in a system relevant to one or more of its stakeholders."

Domains	
Advertising	Entertainment/sports
Aerospace	Environmental monitoring
Agriculture	Financial services
Buildings	Healthcare
Cities	Infrastructure (communications, power, water)
Communities	Leisure
Consumer	Manufacturing
Defense	Science
Disaster resilience	Social networks
Education	Supply chain/retail
Emergency response	Transportation
Energy	Weather

Figure 2.9   From NIST 1500-201.

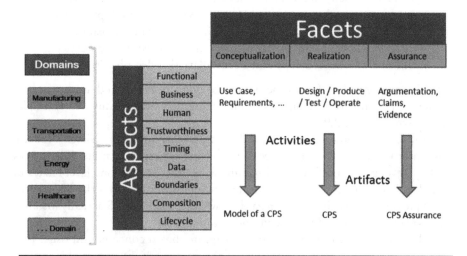

Figure 2.10   From NIST 1500-201.

A lengthy list of concerns in the document provides a rich source of vocabulary that can be used directly, or modified as necessary, to provide the basis of a classification and categorization system for IoT devices in your institution or city.

A partial list of concerns includes:

• Communication	• Safety
• Controllability	• Time awareness
• Functionality	• Data semantics
• *Manageability*	• Operations on data
• Measurability	• Identity
• Monitorability	• Data relationships
• Performance	• Data volume and velocity
• Physical properties	• Networkability
• Sensing	• Adaptability
• States	• Complexity
• Uncertainty	• Discoverability
• Cost	• Deployability
• Quality	• Disposability
• Regulatory	• Maintainability
• Usability	• Operability
• Privacy	• Procureability
• Reliability	
• Resilience	

Another approach is from the publication, *Internet of Things: a definition & taxonomy*, Dorsemaine et al. provides top-level language that can be used for classification and categorization.[17] This work starts with the central concept of a *connected object*. In turn, a connected object can have these characteristics or attributes as well as sub-attributes. These hierarchies can then subsequently be used for classification and categorization. These classifications and categorizations can subsequently be used as a basis for developing "risk buckets" for an institution or city.

■ Energy
  – Energy source. How does the object (device) get its power? Does the power source need to be recharged or replaced on some schedule? Is it powered from local building or similar power? Does it harvest energy by gathering from the environment, such as solar panels or harvesting radio frequency for power?[18] Other?
  – Energy management. Is the object normally off, low power, or always on?
■ Communication
  – Does the communication interface use wired or wireless protocols? Or both?
  – Does the device use local "pickup points" or gateways that connect the object to the network?
  – Is the communication method of the object disconnectable?

- How is communication initiated? By the object? By the gateway? Either?
- How does the device authenticate itself? (Does the device authenticate itself?)
- Is there encryption? What type of encryption?
- Does the device support error correction to address communication disruption?
- What is the bandwidth of the device? What is the maximum rate of data transmission?
- What is the maximum range for the data transmission? Inches? Feet? Tens of feet? More?

■ Functional attributes
- Interactions
  • Does the sensor on the object/device have local memory? Can it save locally obtained values temporarily?
  • Or does the device have to communicate immediately upon sensing?
  • Does the device have an actuator? That is, does it change the local environment in some way – e.g., move something, heat/cool something, light something, change air quality, etc.?
  • Does the device have both sensor and actuator?
- Mobility
  • Is the object/device fixed or does it move?
- Management
  • Can the device be managed? Can the configuration be changed after installation?

■ Local user interface
- Is there a direct user interface?
- Is it active such as buttons? Or passive such as through lights, sounds, or others? Or both?

■ Hardware and software resources
- How much RAM and nonvolatile storage does the device have?
- What kind of processing power does the CPU have?
- Security
  • Hardware encryption onboard?
  • Has the onboard code been proven? Is its behavior deterministic?
  • How does the device identify and authenticate itself?
  • Does the hardware provide the capability to manage its resources or processes?
  • Accountability. Does the device log its own activity?
- Dependability
  • Are there mechanisms in place for ensuring, or increasing the probability, that what the device is sensing *and* reporting are the same. For example, is there redundancy built into aspects of the device.
- Operating system (OS)

- • Does the device have a definitive OS?
- • Is the OS both hardware and software?
- • Or is the logic purely in hardware such as a programmable logic controller?[19]
- – Can the system update its software or firmware?

Another Aspect of Variety and Variation – Variation and Impacts to Supply Chain Risk Management

The second part of the Variety and Variation topic is an aspect that directly impacts supply chain risk management. This has IoT systems manageability impacts in terms of both potentially substantially increased risk to the institution or city, and resourcing, labor, and costs in an effort to mitigate the risk.

> We'd also like to stress that the supply chain attacks will continue to become easier and more prevalent as emerging technologies increase the attack surface exponentially over the years.
>
> Jennifer Bisceglie testimony, US–China Economic and Security Review Commission hearing[20]

Part of this risk simply comes from sheer math – because the number of devices – the T in IoT – is increasing exponentially, the number of subcomponents within those devices is also increasing exponentially – and that number of subcomponents is a lot bigger than the number of IoT devices. Additional substantial risk comes in because, at the time of purchase and deployment, *we don't know where most of those subcomponents come from*. Though some components could source from a known, vetted manufacturer or software developer, many – likely most – will be from other sources. Examples of IoT device component sources includes but is not limited to

- ■ A small, new, unvetted domestic firm with limited production and support history
- ■ A firm from a hostile country or nation-state.
- ■ A firm from an international competitor
- ■ A firm backed or hosted by national or international criminal actors
- ■ Someone's garage
- ■ Other

Recall that the IoT device – the Thing in Internet of Things – is a device that

- ■ Computes
- ■ Is networked
- ■ Interacts with the environment in some way.

To enable these capabilities and to add business value, each device is comprised of many different subcomponents which work in concert to create a functional "thing" or device (Figure 2.11).

Hardware: Because an IoT device interacts with the environment in some way, it has a hardware component or components that serve as a sensor or actuator. There could be more than one of each on a device. A small sample of sensor manufacturers includes:

- NXP/Qualcomm (Netherlands)
- Texas Instruments (USA)
- ST Microelectronics (Switzerland)
- Wulian (China)
- Many others.

Other technologies have similar issues,[21] e.g., the iPhone 7 has a

- Japanese camera
- South Korean memory
- A British power management chip
- Taiwanese wireless hardware
- A Dutch user interface processor
- An American radio.

Hardware driver: A piece of firmware or software that interacts with the sensor or actuator. This may or may not be written by the hardware manufacturer.

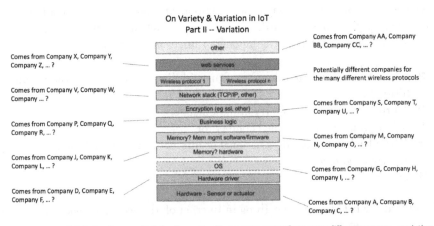

Figure 2.11   Component provenance very difficult to track for IoT devices.

Operating System: The device may or may not have an OS. If it does, it could come from a variety of IoT OS providers. Examples include:

- RIOT OS (open source)
- WindRiver VxWorks (USA)
- Google Brillo (USA)
- ARM mbed (UK)
- Alibaba's AliOS (China).

Memory and memory management software: There may be separate or integrated memory on the IoT device as well as separate or integrated memory management software. Any of these could have multiple sources of development and production.

A snippet from the author's testimony at the US–China Economic and Security Review Commission March 8, 2018, at the "China, United States, and Next Generation Connectivity" hearing[20] relays an anecdote regarding unclear provenance of electronic components:

> While the following anecdote is certainly not indicative of all manufacturing processes, it has always stuck in my mind as a reminder that not everything, i.e., electronic component, may be where I think it's from or coded the way I think it's coded.
>
> Andrew ("bunnie") Huang, MIT electrical engineering PhD, and his business partner Sean ("xobs") Cross xxiii gave a talk at the 2013 Chaos Computer Congress xxiv on hacking SD cards. SD cards are the removable memory cards that go into digital cameras and other electronics.
>
> In the course of the presentation, Huang describes vast bins of memory cards of ranging quality, size, and performance in the market of Huanqiangbei in Shenzhen, China. He talks about card relabeling as a common practice to adjust for sub-performing cards as well as card factories that have very few access controls regarding what configuration files are written to cards and chips and how they are configured. Transcribing from the presentation video (at approximately 50:45):
>
> ... when we've been to the factories where they burn [program] the firmware in, you can basically just walk in and go up to the burner [component programmer] and replace the files on it ... literally, there were chickens running through the factory ... there's no security, there's no badges ... they make these things [components] and ship them all over the world ...
>
> My previous naïve assumption that all electronic parts were created and programmed in carefully controlled and audited environments was appropriately debunked. Many buy from this kind of loosely controlled electronics market because the components are very inexpensive compared to a highly regulated manufacturer. IoT devices have many of these kinds of components.

Business logic: Software encoded business logic. This is the software that does specific unique things with the sensor or actuator and data from and to the same. It can include some local data processing and analytics.

Encryption logic: While computational constraints on many IoT devices will limit the encryption on many devices, at least for the near future, the encryption that is deployed could have a variety of sources to include open source approaches. NIST's "Report on Lightweight Cryptography" (NISTIR 8114) discusses a number of different encryption-related algorithms such as CLEFIA,[22] LED,[23] and Piccolo.[24] These and other algorithms and protocols could be written by any number of different software developers.

Network stack: While there are challenges to implementing traditional TCP/IP networking stacks in IoT devices, we can continue to expect to see them for some time. And like the other software components, these network stacks can come from multiple different providers.

Wireless protocols: There are a large number of wireless protocols being developed and deployed in IoT systems. Some examples[25] include:

- 6LoWPAN[26]
- Bluetooth
- Bluetooth Low Energy[27]
- Zigbee[26]
- LoRaWAN[28]
- NB-IoT[29]
- 3G and 4G
- 5G[30]
- LTE Cat 0,1[31]
- RFID
- Sigfox[32]
- Many others.

All of these protocols could be written by a number of different developers.

Web services: Many IoT devices have built in web servers so that they are directly interactable by humans, other devices, the cloud, and more. Examples of companies writing web servers for IoT devices include:

- Lantronix XChip (hardware and software)[33]
- Allegro RomPager AE[34]
- GoAhead by EmbedThis[35]
- Mongoose by Cesanta[36]
- Others.

So, *each type* of IoT device can have *many different permutations* of component combinations.

Alleles in biology

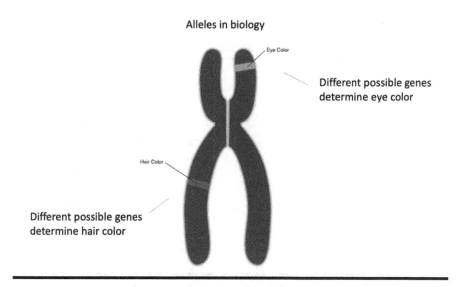

Figure 2.12 **Different genetic strings positioned at particular sections of the chromosome determine particular features. (Image courtesy of Tom Benson.)**

Alleles for IoT Devices

This is not terribly unlike the concept of alleles in genomics. Alleles are the possibilities of a particular gene in a particular location on the chromosome. So, for example, there's an allele for hair color, an allele for eye color, an allele for height in pea plants, and so on. The allele is one of the instances, the possibilities, of what a particular trait – eye color, height, blood type – of a particular gene (Figure 2.12).

So, for an IoT device, say five companies make five different products for IoT devices web services. The code and logic for this web service exist in some logical (and physical) aspect of the IoT device. To use the biology metaphor, each of those possible products that would go into that IoT device to perform a particular function (web services in this case) would be an allele for that function.

It could be helpful to use this analogy of alleles for types of the different component possibilities in IoT devices as we wrestle with the magnitude of

- ■ High numbers of devices
- ■ Even higher numbers of device subcomponents
- ■ Number of device permutations that a type of IoT device can have.

And all of the above – devices, subcomponents, and permutations – are accelerating their infiltration of our campuses, institutions, and cities (Figure 2.13).

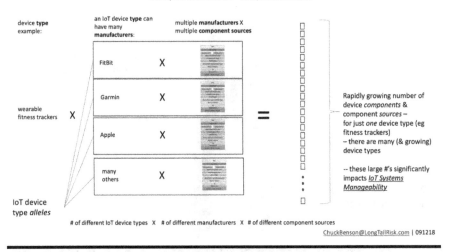

Figure 2.13   IoT component 'alleles' for wearable fitness devices.

This allele idea – that is many permutations of components among IoT device types – has two impacts on IoT system manageability:

■ Another very large number of, and growing, things to deal with within our cities, campuses, and institutions
■ The wide variety of types of subcomponents means that a large number of different active software and hardware subcomponents – *of which we know very little about* – are entering our campuses and cities (Figure 2.14).

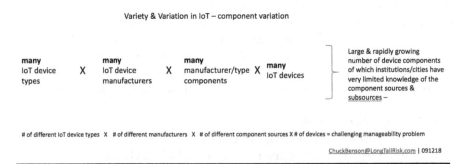

Figure 2.14   Very large numbers of different IoT device components.

Big Numbers and Increased Variety and Variation Change the Game

The number of things to be managed and risk-managed in our city and institutional IoT systems is growing rapidly. Because of this, our traditional approaches to managing technology systems are insufficient and ineffective. Similarly, managing risk around those systems in the traditional approach will not scale well. We are going to need to count fewer things – and greatly lower expectations of traditional risk management and accept that even larger uncertainty. Or, we are going to have to find a new approach to managing this sort of risk, though there will still likely be increased levels of uncertainty from which we are currently familiar.

Suggested Reading

1. Evans, Dave. How the Next Evolution of the Internet Is Changing Everything, 2011, 11.
2. Moomonitor+. *Dairymaster* (blog). Accessed January 8, 2019. https://dairymaster.com/products/moomonitor/.
3. Drone Defense – Powered by IoT – Is Now a Thing | Network World. Accessed January 8, 2019. https://networkworld.com/article/3309413/internet-of-things/drone-defense-powered-by-iot-is-now-a-thing.html.
4. Kastrenakes, Jacob. Kuvée Is Trying to Reinvent Wine with a Ridiculous Wi-Fi Bottle. The Verge, March 28, 2016. https://theverge.com/2016/3/28/11317518/kuvee-bottle-keep-wine-fresh-smart-wi-fi.
5. silicon. 15 Ways American Football Is Tackling the High-Tech Field. *Silicon Republic*, February 3, 2017. https://siliconrepublic.com/machines/american-football-nfl-iot-tech.
6. Internet of Things Device Security and Supply Chain Management. Wilson Center, November 22, 2017. https://wilsoncenter.org/publication/internet-things-device-security-and-supply-chain-management.
7. Kenton, Will. Bucket. Investopedia. Accessed January 9, 2019. https://investopedia.com/terms/b/bucket.asp.
8. *Law of Large Numbers.* Accessed January 9, 2019. https://khanacademy.org/math/statistics-probability/random-variables-stats-library/expected-value-lib/v/law-of-large-numbers.
9. Bernstein, Peter L. *Against the Gods: The Remarkable Story of Risk.* New York, NY: Wiley, 1998.
10. The Worst Form of Government. The International Churchill Society, February 25, 2016. https://winstonchurchill.org/resources/quotes/the-worst-form-of-government/.
11. Benson, Chuck. Creating Initial IoT Risk Categories. *Long Tail Risk* (blog), June 12, 2015. http://longtailrisk.com/2015/06/12/creating-initial-iot-risk-categories/.
12. Nest Cam Indoor | This Is What a Home Security Camera Should Be | Nest. Accessed January 10, 2019. https://nest.com/cameras/nest-cam-indoor/overview/.

13. Jin, Ming, Nikolaos Bekiaris-Liberis, Kevin Weekly, Costas Spanos, and Alexandre Bayen. Sensing by proxy: Occupancy detection based on indoor CO_2 concentration, n.d., 10.
14. The Best Access Control Systems of 2019. business.com. Accessed January 10, 2019. https://business.com/categories/best-access-control-systems/.
15. Griffor, Edward R, Chris Greer, David A Wollman, and Martin J Burns. *Framework for Cyber-Physical Systems: Volume 1, Overview.* Gaithersburg, MD: National Institute of Standards and Technology, 2017. doi:10.6028/NIST.SP.1500-201.
16. 14:00–17:00. ISO/IEC/IEEE 42010:2011. ISO. Accessed January 11, 2019. http://iso.org/cms/render/live/en/sites/isoorg/contents/data/standard/05/05/50508.html.
17. Dorsemaine, Bruno, Jean-Philippe Gaulier, Jean-Philippe Wary, Nizar Kheir, and Pascal Urien. Internet of Things: A Definition & Taxonomy, 2015. doi:10.1109/NGMAST.2015.71.
18. Harvesting Energy from RF Sources. Microwaves & Radio Frequency, December 30, 2016. https://mwrf.com/systems/harvesting-energy-rf-sources.
19. ProgrammableLogicController.In *Wikipedia*,December30,2018.https://en.wikipedia.org/w/index.php?title=Programmable_logic_controller&oldid=875964590.
20. Cleveland, Robin, and Carolyn Bartholomew. Hearing on China, the United States, and Next Generation Connectivity. U.S.-CHINA ECONOMIC AND SECURITY REVIEW COMMISSION, 2018, 132.
21. Ip, Greg. Bringing IPhone assembly to U.S. would be a hollow victory for trump. *Wall Street Journal*, September 19, 2018. Sec. Economy. https://wsj.com/articles/bringing-iphone-assembly-to-u-s-would-be-a-hollow-victory-for-trump-1537368671.
22. Shirai, Taizo, Kyoji Shibutani, Toru Akishita, Shiho Moriai, and Tetsu Iwata. The 128-bit blockcipher CLEFIA (Extended Abstract). In *Fast Software Encryption*, edited by Alex Biryukov, 181–95. Lecture Notes in Computer Science. Berlin: Springer, 2007.
23. Guo, Jian, Thomas Peyrin, Axel Poschmann, and Matt Robshaw. The LED block cipher. In *Cryptographic Hardware and Embedded Systems – CHES 2011*, edited by Bart Preneel and Tsuyoshi Takagi, 326–41. Lecture Notes in Computer Science. Berlin: Springer, 2011.
24. Shibutani, Kyoji, Takanori Isobe, Harunaga Hiwatari, Atsushi Mitsuda, Toru Akishita, and Taizo Shirai. Piccolo: An ultra-lightweight blockcipher. In *Cryptographic Hardware and Embedded Systems – CHES 2011*, edited by Bart Preneel and Tsuyoshi Takagi, 342–57. Lecture Notes in Computer Science. Berlin: Springer, 2011.
25. Chaudhuri, Abhik. *Internet of Things, for Things, and by Things.* Boca Raton, FL: CRC Press/Taylor & Francis Group, 2019.
26. Ray, Brian. 6LoWPAN vs. ZigBee: Two Wireless Technologies Explained. Accessed January 13, 2019. https://link-labs.com/blog/6lowpan-vs-zigbee.
27. The Basics of Bluetooth Low Energy (BLE). EDN. Accessed January 13, 2019. https://edn.com/5G/4442859/The-basics-of-Bluetooth-Low-Energy--BLE--.
28. About LoRaWANTM | LoRa AllianceTM. Accessed January 13, 2019. https://lora-alliance.org/about-lorawan.
29. Narrowband – Internet of Things (NB-IoT). *Internet of Things* (blog). Accessed January 13, 2019. https://gsma.com/iot/narrow-band-internet-of-things-nb-iot/.
30. phones, Mike Moore 4 days ago Mobile. What Is 5G? Everything You Need to Know. TechRadar. Accessed January 13, 2019. https://techradar.com/news/what-is-5g-everything-you-need-to-know.

31. LTE Cat0 for M2M | LTE-M Rel Cat-0 Cat-1 Cat-4. Accessed January 13, 2019. http://rfwireless-world.com/Terminology/LTE-Cat0-for-M2M-LTE-M.html.

32. Sigfox Technology Overview | Sigfox. Accessed January 13, 2019. https://sigfox.com/en/sigfox-iot-technology-overview.

33. XChip | Lantronix. Accessed January 27, 2019. https://lantronix.com/products/xchip/.

34. RomPager AETM. *Allegro Software Development Corporation - Secure Software for the Internet of Things* (blog). Accessed January 27, 2019. https://allegrosoft.com/embedded-web-server-ae.

35. Embedded Web Server - GoAhead | Embedded Web Servers | Embed this Software. Accessed January 27, 2019. https://embedthis.com/goahead/.

36. Mongoose OS - Licensing Information. Accessed December 10, 2018. https://cesanta.com/comparison.html.

Chapter 3

Defining IoT Systems Implementation Success

A successful Internet of Things (IoT) systems implementation has two overarching components:

- Return on Investment (ROI)
- Changes to the institution's cyber risk profile.

Both of these components can be challenging to determine.

ROI for IoT

Calculating ROI can be deceptively difficult to calculate for IoT systems. It involves determining the actual benefit received or value added, particularly as compared to the expectation of the benefit or value, less the cost of the initial investment plus ongoing support costs which include hardware, software, and most importantly a number of different types of labor.

Data Expectations as Part of ROI Calculation

Calculating the benefit received, the value added or created, is difficult because the many different constituencies across a city or institution may have different interpretations and perspectives of value. This is particularly true around expectations of data. Data creation and analysis is often the ostensible primary benefit of IoT systems.

For example, an air quality sensing system with sensors distributed geographically across a city could have different performance expectations for different groups of people. In most cases, the bulk of that performance expectation may be in the form of expectations and interpretations of data and, most importantly, actionable data.

Various interests and expectations can vary substantially by the role or population using the data. Some hypothetical examples are

Role:	*Interests Could Include*
City manager	A dashboard of daily updates of air quality data
Term-based elected official (e.g., mayor, council person)	A short-term trend, e.g., <4 years, of air quality improvement or degradation
University researcher	A longer-term trend, e.g., one to three decades, of air quality improvement or degradation
Compliance organization (local, state, fed)	Data on specific particulates as snapshot in time and time studies that may collect multiple snapshots over time
City chamber of commerce	Marketable air quality statistics, e.g., in comparison with other cities
University researcher	Observed data compared with human survey data
Voting, vocal citizen in geographic region X	Personal perception of air quality and quality of life
Voting, vocal citizen in geographic region Y	Personal perception of air quality and quality of life
Voting, vocal citizen in geographic region Z	Personal perception of air quality and quality of life
Operations support staff and planners	Information on determining impact of city operations (vehicles, power production, steam production, etc.) on air quality

Another example regarding how different roles and constituencies within a city or institution have different expectations of technology and the data it produces might be a networked video surveillance system deployed across the city is tabulated as

Role:	Interests Could Include
Law enforcement	Identification of malicious activity for prevention/intervention. Evidence for prosecution
City planner	Traffic pattern and volume changes over time – pedestrian, vehicles, bikes, other
Parks planner	Parks usage over time – equipment, fields, lights, restrooms, etc.
Parking enforcement	License plate recognition technology. Automatic Vehicle Identification (AVI)
Road tolling	AVI
Voting, vocal citizens group	Privacy concerns – what is collected and where and used for what purpose

These examples are just a small subset of possible interests, interpretations, and expectations of an IoT system's data output. How that IoT system performs could have many interpretations that vary with the individual, group, or organization reviewing the data.

Because of these many possible expectations of data, determining the value add or return can be particularly challenging.

Researchers Fiore-Gartland/Neff[1] have proposed a framework for considering those data expectations in the context of health and wellness data from which we might borrow in considering IoT systems data in institutions and campuses in which the concept of data valences is introduced.[2] In this context, the term "valence" is more closely aligned with Merriam-Webster's second definition of "relative capacity to unite, react, or interact."[3]

The authors identify six data valences:

■ Self-evidence
■ Actionability
■ Connection
■ Transparency
■ "Truthiness"
■ Discovery.

I'll briefly describe these valences within the healthcare context in which they were developed and then suggest how they might be applied to an IoT system in an institution or city such as an energy management or smart grid system.

Self-Evidence Valence

The idea of the self-evidence valence is that data is context-free or at least appears that way. The context-free-ness notion conflicts with the popular assumption of interpretation or mediation being required to make data meaningful as the researchers point out. I believe that data does indeed need mediation to be relevant. Data without mediation tends to devolve to the unsatisfying "just because" answer.

Actionability Valence

Actionability refers to the expectation that the data does something, directly or indirectly. From the context of the data consumer, can that data be used to do something meaningful for that consumer within their context? The example is given of a physician being presented with self-collected patient data. This may well not be "clinically actionable" because the physician has no basis for comparison or reference.

Connection Valence

This valence identifies data as a "site for conversation." (p. 1475) This is a particularly powerful valence because the connection valence draws people to the same table, whether figurative or literal, to discuss data for one reason or another. One example is that of a home patient contacting their case manager about data being collected as a part of the telemedicine system. While the call was not particularly important regarding the telemedicine question, it did provide an opportunity for the case manager and patient to connect and share other information which may have been recorded in an informal way.

Even if the data-discussion reasons are simple or seem unrelated to ostensible objectives, people are still showing up, meeting, and talking for whatever reason and in the course of that showing up, other things are shared and communicated. This is a powerful valence and yet not easily quantifiable.

Transparency Valence

The transparency data valence is the idea or expectation of real or perceived benefit of data being "accessible, open, sharable, or comparable across multiple contexts …. Making data transparent across communities is one set of values or expectations." (p. 1475) The transparency valence also introduces the idea that when there is data transparency, when it is indeed shared across contexts, new questions around ownership, access, and confidentiality present themselves. Addressing these new

questions/issues of data ownership, data access, and data confidentiality is important work and requires institutional resources in the form of staffing and skill sets
that have traditionally not been planned for or organic to the institution.

Truthiness Valence

Stephen Colbert popularized the word "truthiness"[4] during one of his shows.
He uses truthiness to describe something that feels right or just seems right, generally without regard to facts or evidence. Similarly, the truthiness data valence, in
turn, has to do with the data quality of the data simply feeling right or seeming
right and without reference or supportive logical argument. The concept of truthiness might also have more serious parallels to Ockham's razor, where the simpler
explanation is preferred to the more exhaustive and/or complex.[5]

Discovery Valence

The discovery valence "describes how people expect data to be the source or site
of discovery of an otherwise obscured phenomenon, issue, relationship, or state."
(p. 1477) This is inconsistent with the popular notion of Big Data where often broad
assumptions are made that simply because there is an unprecedented amount of data,
that there must be new patterns, knowledge, and possibly unsolved mysteries there.

Applying the Data Valence Concept to IoT Systems – An Energy Management Example

We can use the illustration of an institutional energy management system. This is
an IoT system consisting of hundreds or thousands or more of networked meters
and building automation systems sensors and actuators such as that often found on
a Higher Education campus or within a city. Below is one approach on how each of
these valences might come into play in this context.

Self-Evidence Valence in Energy Management Data

I don't see this valence playing out particularly well in the IoT systems space. This
energy management data sourced from thousands of energy sensors across a city
or institution needs to have context and be interpreted to have relevance. Also, the
data is too new, unfamiliar, sometimes abstract, and often complex for there to be
strong statements of self-evidence.

That said, the topic of climate change and all the misinterpretations, rheto-
ric, and claims (from either side of the argument) made without context comes to
mind. So maybe the self-evidence valence has applicability here as well. Perhaps,
conclusions will indeed be drawn from energy data devoid of context.

Actionability Valence in Energy Management Data

This valence definitely applies to the energy management example as well as IoT systems more generally. Indeed, actionable data – or lack thereof – is a key aspect of discerning the value or utility of an IoT system. Everyone – consumers, vendors, government, others – expects to do something useful – save money, reduce carbon, perform better institutional or city planning – with energy management data collected via IoT systems. From an interview with a program manager at a large building automation and energy management services provider:

> Data must have context. That context usually comes from more static information (place, type, procedure, design intent...). The live data is only actionable when there is contextual data that helps you define thresholds or alarms and then helps you actually resolve. The hard work is in the middle with workflow design and management ... live data is only of value if you have foundational ... data to go with it.[6]

Connection Valence in Energy Management Data

This valence is present in energy management systems as well as other IoT systems. This data provides the place, or the kernel, to come together to problem-solve. In the course of that problem-solving, a parade of assumptions and expectations come quickly to the surface. Finance professionals, energy management professionals, IT and data professionals, vendors, and a variety of end users bring their expectations, assumptions, and desires to these meetings, Skype calls, phone calls, and emails.

This data valence is particularly important at this point in the respective evolutions of energy management systems, IoT systems in general, smart campuses and smart cities. Facilitating the opportunity for city leaders, citizenry, and systems providers to come together to problem-solve is critical to success.

Transparency Valence in Energy Management Data

Everybody (almost) wants transparency of data in energy systems and most other IoT systems purchased by an institution or city. This distribution of data interpretation across contexts is exciting and challenging, and also fraught with peril for misunderstanding. That said, addressing topics around this valence can bring important issues to the surface.

Truthiness in Energy Management Data

I believe the "truthiness" valence does have some play here in energy management systems as well as other IoT systems. It is not particularly unlikely for cases of confirmation bias to occur where new and unfamiliar data is interpreted to align with preconceived notions such as, "We installed new solar panels in the institution, therefore we must be saving money," ... or "We implemented a complex new building automation system, so we must be more efficient."

Discovery Valence in Energy Management Data

There is a high expectation for this valence in energy management systems. That said, while the expectation of discovery is high for energy management data, it is often harder to actually attain – often requiring more staffing, skill set, and time – than originally anticipated.

Data Valences in IoT Systems

How we, across our multiple constituencies within a city or institution, perceive various aspects of data has a strong influence on the perceived success of the system that produced the data. That perceived success, in turn, directly impacts the *Return* aspect of the ROI criteria of measuring IoT system success. This is true for city energy management systems, and I believe that that is broadly generalizable to IoT systems of institutions and cities.

Understanding data perceptions across an institution or city is essential for successful IoT system implementations. The capability and capacity of a city or institution to implement complex IoT systems in a complex environment is essential to success. Understanding the varied data consumers and their perceptions and needs in a complex organization such as a city or campus is, in turn, a critical component to a successful IoT system implementation.

Determining Actual Costs of Systems Deployments

Once expectations are established and defined across possibly several groups or constituencies, creating a basis of value, some analysis can be done on whether that value has been reached – which is the Return part of the ROI equation.

The next step in determining ROI is discerning the actual costs incurred in implementing and operating the IoT system. These costs can be deceptively high because of the large number of devices deployed, the geographical distribution of these devices, the professional skill sets needed to support devices in these deployments, and the risks and liabilities incurred in deploying these devices. This is all in addition to the more traditional IT components of server application implementations, client desktop applications, and even "traditional" cloud services (as traditional as one or two decades can be). So, the number of costs, the types of costs, the institutional organizations bearing these costs, and the novelty or unfamiliarity of these costs (e.g., the organization bearing the cost wasn't expecting it) all combine to make costing IoT systems challenging and labor-intensive.

Determining the total cost of ownership, operation, and stewardship of IoT systems for an institution or city has a number of considerations.[7] Some of these considerations are shared with traditional enterprise systems, and some are unique to IoT systems. Lack of realization, acknowledgement, and management of these costs can lead to disappointing and unnecessarily costly IoT systems outcomes. These disappointing outcomes manifest themselves in the lost ROI of an IoT investment as well as negative impacts to the cyber risk profile to an institution.

We can group costing areas broadly and list from least complex and nuanced to more complex and nuanced:

1. Costs to host servers, databases, and provide redundancy – whether under desks (hopefully not), in local data centers, or in the cloud
2. Costs to support large numbers of geographically distributed "Things" and devices (the T in IoT), and the different institutional organizations that may be involved (some of which may not see or realize the potential benefit to the institution but may bear some portion of its support costs)
3. Costs stemming from the natural friction, sometimes small and sometimes large, between multiple historically disparate organizations within an institution as they attempt to coordinate, collaborate, and address the *problems of understanding*[8] to support the system. These costs, aka system loss, will be discussed more fully in Chapter 5.

Costing IoT Systems – the Easier Part

While this type of costing is more traditional and has been done before, it still requires professional skill sets, time, and can be tedious.

This costing is more closely aligned with traditional enterprise application costing than other IoT systems costs. Application licensing and support agreements are a part of this aspect. Important additional costs to include, though, are hosting and support costs for those applications and databases. Also, someone has to manage that hosting with the provider as a part of their job description. What are the costs of that management, e.g., who manages the relationships, the tickets, the problems, etc.?

The applications servers or virtual machines (VMs), database servers/VMs, and redundancy servers/VMs can be hosted in a local data center, shared data center, the cloud, or similar. Hosting or otherwise servicing these applications and supporting databases have their own costs. In addition to fees for the hosting, there's also some cost to managing the vendor relationship and agreement/contract.

Costing items in this area includes such questions as

■ Is there an aggregating data server or servers (and/or device controlling servers)?
  – Is on-premise hardware or VM?

- What are the costs of these?
- *Who* pays for this?
  - ■ Does the vendor offset these costs in some way? Or does the buyer receive the full load of the costs?
- Are there hosted hardware/VM hosting costs? For example, internal institutional or city costs to install and support the hardware and/or VM?
  - ■ *Who* pays for this?

■ Are separate databases needed to support the application?
  - How many?
  - Database licensing cost?
  - Hardware supporting database costs?
    - Whether actual hardware or through for-fee services such Amazon Web Services (AWS) database services?[9]
  - Can database requirements be met with an existing database (without degrading the existing database performance)?

■ Is there an aggregating data service (and/or device controlling service), e.g., cloud-based/Software-as-a-Service?
  - What is the cost of this/these service/services?
  - *Who* pays for this?
    - Does the vendor offset these costs in some way? Or does the buyer receive the full load of the costs?

■ Data aggregation software/application
  - Regardless of where it is run, e.g., on-premise or as a service:
    - What is the original purchase cost?
    - What are the ongoing support costs?

■ Supporting client applications (running on workstations, laptops, tablets, smart phones, etc.)
  - Who pays for these devices?
  - If client application is packaged independently of centralized data aggregation/device control application:
    - What is client application cost?
    - How is it priced? Per named user, concurrent users, etc.?
    - What are annual support and maintenance costs?
    - How many licenses need to be purchased?

■ Redundancy/disaster recovery/failover
  - Are clustered servers or services available?
    - For application? For database?
    - What are the costs?

■ Is redundant hardware needed?
  - What are the costs?

■ Who will manage the contractual aspects of all of these components?
  - How much time effort will be required?

- How many hours? Percentage (%) FTE?
- Is there capacity to support this? Will it disrupt existing work?
- What are the effects of insufficient support capacity?
    - To existing systems
    - To the new system.

These are all fairly typical questions that analysis of an enterprise IT system would include. The following section, that considers the newer and organizational boundary spanning aspects of the endpoint devices of IoT systems, steps up the complexity.

The Less Familiar and More Nuanced Part of IoT Systems Costing – IoT Devices

I call this "the guy, a truck, and a ladder problem" (with the "guy" being a female or male skilled professional). There is considerable cost to deploying, maintaining, replacing IoT endpoints – and there are *many* endpoints/devices (Figure 3.1).

Imagine an institution or city that implements 500 smart streetlights that provide lighting, sense movement, maybe sample air quality, and possibly monitor and report street or underground vibration. There will be some failure rate among the components in any single device/endpoint and some failure rate across the total number of installed devices/sensors/actuators – the T's in IoT.

In this hypothetical scenario, troubleshooting and/or repair of a single device deployed in some location in the field in the institution or city – and likely requiring

Figure 3.1 This is not cheap – skilled staff, vehicles, equipment, risk (hazard), and others, ... and this is just one device of potentially hundreds, thousands, ...

a ladder (to go high, e.g., a sensor on a streetlight – or to go low, e.g., to lower the professional into a hole, tunnel, or other) – or be otherwise embedded in the environment such as a ground-level sensor means:

- Deploying one or more skilled tradespeople
  - One or two for the work and possibly another as a safety observer for a total of at least two to three skilled tradespeople.
- Rolling a truck or trucks
  - With associated vehicle maintenance, fuel, costs, insurance and other costs.
- Spending 1–2 hours preparing for the troubleshooting/repair
  - Reviewing work order, assembling tools and parts, travel, etc.
- 2–4-hours troubleshooting/repair
- 1–2-hours wrap-up and return.
  - Site cleanup and securing, completing documentation/work orders, return travel, tool and parts return, etc.

For this hypothetical example, let's say the skilled tradespeople involved make $60/hour and their benefit load (expense to institution or city) is 25% for a total $75/hour expense. So, disregarding fleet and related costs, one estimate is shown in Figure 3.2.

Continuing on the thread, let's say there's a 10%/year failure rate (or required hardware/software update rate) of at least some component on a single device or Thing (the T in IoT). That is shown in Figure 3.3.

That dollar figure of total annual support cost of that system starts to become nontrivial. And that's just *one* IoT system. Cities and institutions will have portfolios of many of IoT systems managed to varying degrees of effectiveness (and with increasing interdependence between them). Another hypothetical example is shown in Figure 3.4.

Figure 3.2 Calculating support event costs. Substitute costs that make sense for your institution.

Figure 3.3 Estimating annual support costs for a device in an IoT system.

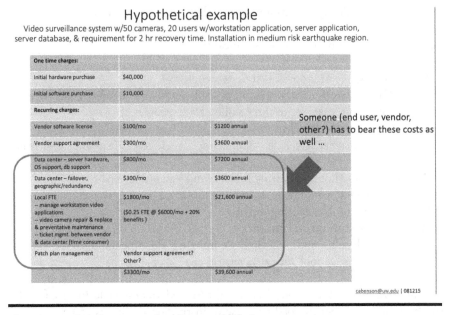

Hypothetical example

Video surveillance system w/50 cameras, 20 users w/workstation application, server application, server database, & requirement for 2 hr recovery time. Installation in medium risk earthquake region.

One time charges:			
Initial hardware purchase	$40,000		
Initial software purchase	$10,000		
Recurring charges:			
Vendor software license	$100/mo	$1200 annual	Someone (end user, vendor, other?) has to bear these costs as well ...
Vendor support agreement	$300/mo	$3600 annual	
Data center – server hardware, OS support, db support	$800/mo	$7200 annual	
Data center – failover, geographic/redundancy	$300/mo	$3600 annual	
Local FTE -- manage workstation video applications -- video camera repair & replace & preventative maintenance -- ticket mgmt. between vendor & data center (time consumer)	$1800/mo ($0.25 FTE @ $6000/mo + 20% benefits)	$21,600 annual	
Patch plan management	Vendor support agreement? Other?		
	$3300/mo	$39,600 annual	

cabenson@uw.edu | 081215

Figure 3.4 Another example of cost calculations.

The Insidious Part – Loss from Organizational Friction

The least apparent and possibly most costly aspect is the loss that occurs from the coordination and collaboration required between organizations.[10] The idealized scenario of institutional capacity is that there is a homogeneous and near boundless set of resources that includes components such as available staffing levels (FTE), requisite skill sets (technical, operational, and interpersonal), support funding, political/institutional will to support the implementation and operation of the IoT system, vendor relationships, and others (Figure 3.5).

In practice, however, this institutional capacity is comprised of many different organizations and the interrelationships between them.[10] While collaboration and interorganizational cooperation are typically universally lauded, we all know from personal and professional experience that often collaboration between institutional organizations does not in fact work so well. Research has also been done on this phenomenon where, *the discrepancy between the promise of collaboration and the reality of persistent failure* is studied.[11]

Wherever two or more organizations interact with each other, some sort of friction or system loss is present. In the organizational friction example, losses can come from a multitude of sources where the number of friction sources and the intensity of each can vary from organizational relationship to organizational relationship. Examples of this sort of friction/system loss and resultant loss in expected institutional capacity include (Figure 3.6):

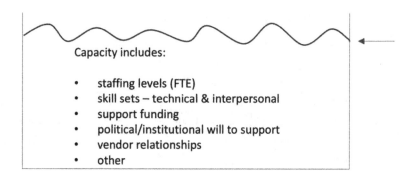

Capacity includes:

- staffing levels (FTE)
- skill sets – technical & interpersonal
- support funding
- political/institutional will to support
- vendor relationships
- other

ChuckBenson@longtailrisk.com | 050917

Figure 3.5   Idealized capacity to manage IoT systems.

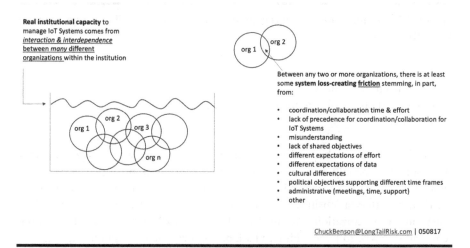

Real institutional capacity to manage IoT Systems comes from *interaction & interdependence* between *many* different organizations within the institution

Between any two or more organizations, there is at least some **system loss-creating** friction stemming, in part, from:

- coordination/collaboration time & effort
- lack of precedence for coordination/collaboration for IoT Systems
- misunderstanding
- lack of shared objectives
- different expectations of effort
- different expectations of data
- cultural differences
- political objectives supporting different time frames
- administrative (meetings, time, support)
- other

ChuckBenson@LongTailRisk.com | 050817

Figure 3.6   Actual capacity stems from multiple organizations with multiple relationships and the friction between.

- Limited administrative resources (planning, meetings, time, support, etc.)
- Lack of precedence for coordination/collaboration for IoT systems, i.e., no patterns to follow
- Misunderstanding/problems of understanding[8] between organizations
- Misalignment of objectives between organizations
- Conflicting obligations to different scopes, projects, and organizations[12]
- Lack of shared objectives
- Different expectations of effort
- Different expectations of data[1]

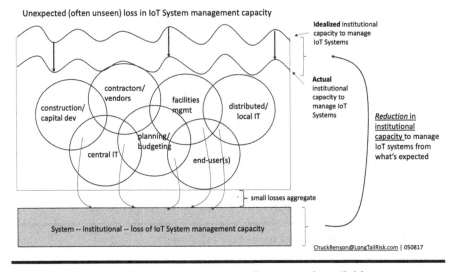

Figure 3.7 **Systems losses aggregate to reduce actual available management capacity.**

- Cultural differences
- Political objectives supporting different time frames[12]
- Others.

The insidious part is that while the friction between any two organizations may be small and possibly not obvious, these small instances of organizational friction aggregate across the whole institutional and IoT system implementation effort. Further, not all relationships are one-to-one. There are often many relationships where many organizations are involved. Likening to Newton's three-body problem,[13] adding a third planet, billiard ball, or organization can significantly increase the complexity of analysis, prediction/forecasting, and the ability to get work done (Figure 3.7).

Costing Differently

In practice, capturing all of these costs can be challenging. Regarding the more traditional technology costing – application and database licensing and hosting costs are relatively the most straightforward and for which we generally have the most experience, but that doesn't mean it's effortless. We have less experience costing the many IoT systems endpoints – the guy, truck, and ladder costs – (guy being female or male) but that support cost can be estimated and the failure rates of devices and device components can be estimated. Finally, the costing that is by far the most challenging and potentially disruptive is the cost-aggregated interorganizational friction. This aggregation of nuanced costs due to systems loss is also possibly the most impactful – in part because of its magnitude and the uncertainty it introduces.

This institutional systems loss and its manifestations of resource depletion, waste, collaboration failure, and others will be discussed further in Chapter 5.

The important thing is to acknowledge all of these components, compute and estimate what we can, and work to allow for and hopefully prepare for the unique uncertainty that selecting, procuring, implementing, and managing IoT systems brings. If we do this work, we have our best chance of reaching the desired ROI.

Assessing Changes in Cyber Risk Profile

The second overarching aspect of assessing the success of an IoT systems implementation is the post-deployment change in cyber risk to the institution or city. Assigning quantitative values[14] to an institution's or city's cyber risk profile is a notoriously difficult and hotly debated endeavor.[15] A good friend of mine and Chief Information Security Officer of a large institution is fond of saying (with great vigor), "There are no metrics in a bar fight!" His point, in part, is that things are so fluid and there is so little common baseline that metrics have limited utility. While some may not express this sentiment as fully, it is generally accepted that establishing strong, actionable risk metrics is difficult. That said, there are things that can be done to look at the delta between risk before and after an IoT systems deployment (Figure 3.8).

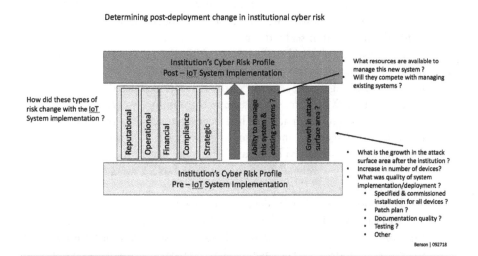

Figure 3.8   Changes to institutional cyber risk.

We can do this by looking at five traditional risk areas,[16] growth in the attack surface area, and ability to manage this system (as well as manage existing IoT systems) based on finite resources. Five fairly traditional risk areas are

- Reputational risk
- Operational risk
- Financial risk
- Compliance risk
- Strategic risk.

An example of what can go wrong is documented in the 2017 Verizon Data Breach report[17] involving an internal attack by IoT devices on a university. The report describes over 5,000 IoT devices, to include lights and vending machines, that attacked the university's Domain Name Servers (DNS), and slowed or disabled substantial portions of the university's network.

Further, we can use the example as part of a basis to develop a guidance document for assessing the real and potential impact of implementing a new IoT system.

From the report:

> Responding to slow or inaccessible network access, the university's help desk and subsequently security team, "found a number of concerns ... The name servers, responsible for Domain Name System (DNS) lookups, were producing high-volume alerts and showed an abnormal number of subdomains related to seafood. As the servers struggled to keep up, legitimate lookups were being dropped – preventing access to the majority of the internet ... From where were these unusual DNS lookups coming? And why were there so many of them? Were students suddenly interested in seafood diners? Unlikely ...
>
> ... firewall analysis identified over 5,000 discrete systems making hundreds of DNS lookups every 15 min ... nearly all systems were found to be living on the segment of the network dedicated to our IoT infrastructure. With a massive campus to monitor, everything from light bulbs to vending machines had been connected to the network for ease of management and improved efficiencies ... While these IoT Systems were supposed to be isolated from the rest of the network, it was clear that they were all configured to use DNS servers in a different subnet.
>
> This botnet spread from device to device by brute-forcing default and weak passwords. Once the password was known, the malware had full control of the device and would check in with command infrastructure for updates and change the device's password – locking us out of 5,000 systems.

This was a mess. Short of replacing every soda machine and lamp post, I was at a loss as to how to remediate the situation. We had known repeatable processes and procedures for replacing infrastructure and application servers, but nothing for an IoT outbreak."

While this example involved several IoT systems – lights, vending machines, others – any one IoT system can be parsed out for risk that it may have (did) add to the institution by implementing an IoT system. Further, while not good for this university, this example also has the benefit of illustrating the additional increased risk that comes from implementing an IoT system poorly, partially, or all of the above. For example, the narrator talks about how the IoT network segment was supposed to be separate from other networks, but it was in fact communicating with DNS servers in other network segments. This poor implementation further exacerbated the increased cyber exposure to the institution created by the addition of the IoT system with limited mitigation planning.

Changes to Reputational Risk

How might this sort of event affect an institution's or city's reputation?

Reputational risk, or brand risk, is at stake when something about the IoT system implementation might cause us embarrassment or loss of confidence in the organization or brand.

Notably, the contributor/narrator in the Verizon report never names the university where this occurs. Surely, this is, at least in part, due to reputation and branding concerns of the university. What kind of negative impact could such an event have to an institution in an ecosystem of universities that are highly competitive for students, faculty, student athletes, parent tuition dollars, donor dollars, state dollars, and others?

- Would this event make the IT organization appear incompetent? The facilities management group appear incompetent? The university administration as a whole appear incompetent?
- Would this event generate privacy concerns for students, faculty, staff, parents, donors, etc.?
- Would this event impact state funding in the next or subsequent budget cycles?
- Would this event make the university appear cavalier to protecting student interests? Donor interests? State/government interests?
- Would this event make federal funding agencies think twice about awarding large research grants to university faculty?
- Would university faculty worry that their ability to get grants would be impacted?
- Would the university suffer negative impact its ability to recruit top performing faculty and graduate students?

■ If the system is maliciously attacked or otherwise compromised, how will this event appear to internal constituents (students, faculty, staff in this case), institutional benefactors (donors in this case), tax payers, prospective students, and faculty?

Changes to Operational Risk

What effects does such an event have on operational aspects?

■ Network performance severely degrades to the point of being unusable
■ With a downed network, there is inability for students, faculty, and staff to perform their learning, research, and job functions
■ The university security team, senior leadership, help desk, and others become absorbed, or at least distracted, by this event and that creates opportunity cost in getting other work done
■ Have safety and/or security issues been created with network outage?

Changes to Financial Risk

What are financial impacts of such an event?

■ Work does not get done
■ Combined with reputation impacts above, there may be a medium- to long-term reduction in revenues – ranging from tuition reduction, state funding reduction, donor dollars reduction, and more
■ Dollars, hours, or both are spent on external consulting resources to help mitigate the problem.

Changes to Compliance and Regulatory Risk

What compliance and regulatory risk might the institution or city face after such an event?

■ Are there mandated services that the institution is unable to perform because of the loss of computing/networking capability or capacity?
■ Is there a failure in mandated controls stemming from the degradation or outage?
■ Are there secondary compliance/regulatory costs due to institutional inability to provide services?
■ Is there a degradation or failure in required oversight in some institutional activities?

Changes to Strategic Risk

What strategic institutional issues are impacted by this sort of event?

- Does this kind of event affect the ability of the institution to execute its mission?
- If so, how long? Short duration? Permanently?

Changes to Attack Surface

The next component to consider when analyzing IoT systems implementation success is what are the changes to the city's or institution's attack surface as a result of the IoT system deployment? Did the institution's or city's cyber risk posture degrade?

For our purposes, the attack surface of an institution, city, or corporation is all of the places that are exposed to attack. It is a list or an array of all of the places that an attacker could get into the institution, city, or business considered in total. The notion of attack surface can also be applied to an application. In this case, the attack surface is all of the places that an attacker could get into the application.

Prior to the IoT phenomenon, the attack surface of a city or institution was largely comprised of all of the servers, desktop computers (and applications on them), laptops, and handheld devices. For example, a single desktop computer with a weak password and out of date antivirus software might offer a point of entry into (1) that particular computer and (2) the organization as a whole. Not every computer will be poorly configured and maintained (hopefully), but some will. The more of these desktop computers that there are in the organization, the more likely it is that there is one or more are exposed – and hence attractive to attackers – computers. And the larger the number of computers, the bigger the chance that there are exposed entry points into the institution, city, or corporation.

IoT substantially exacerbates this problem. Remember, the T in IoT, the Thing in the Internet of Things, is a device that

- Computes
- Is networked
- Interacts with the environment in some way (e.g., sensing or actuating).

These sensors and devices are true fully functional networked computers. It's easy to forget that given the wide variety of (small) form factors and the embedded, non-visible locations of their respective installation sites that they are actually fully functioning networked computers.

The growth rate of these devices appears to be exponential for at least the next few years. Some sources show growth rates of devices of approximately 20% year over year,[18] while others show them as high as 35% year over year (Figure 3.9).[19]

As the increasing number of these IoT systems is deployed with their tens, hundreds, thousands, or more of sensors and actuators, institutions and cities are greatly increasing their respective attack surfaces. Each of these tens, hundreds, thousands, or more of IoT devices is an increased opportunity for exploit and attack. The raw increase in attack points alone broadens the attack surface. The problem is substantially exacerbated though by the high likelihood that those deployments of sensors/devices – candidate attack points – were partially and/or poorly deployed and hence substantially more vulnerable to attack.

A compromised IoT device can be used in many ways. It could be used to disrupt the function of the device itself by blocking data or controllability of the device. For example, a compromised audio/video microphone/camera can have its data streams interrupted and render the device ineffective for its intended use. This is problematic for fairly obvious reasons. Or a compromised energy meter or set of energy meters could stop returning data or, worse, return inaccurate data. Compromised clinical devices such as pacemakers, infusion pumps, glucose monitors, and others speak for themselves regarding negative impact. (The patient using the compromised clinical device could also speak for the negative impact.) These are all clearly important, but there is yet another substantial potential impact from misuse or compromise of IoT devices and systems (Figure 3.10).

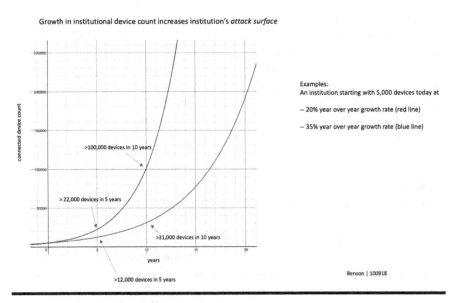

Figure 3.9   **Exponential growth rates.**

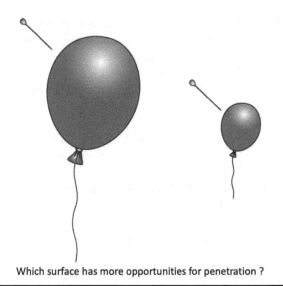

Which surface has more opportunities for penetration ?

Figure 3.10 Growth in area of attack surface represents more opportunities or venues for attack. (Image courtesy of Tom Benson.)

This other critical aspect of exposure is that IoT systems and devices in those systems can be compromised and then repurposed for malicious use with just the computing and networking capabilities of the device. The sensing or actuating portions of the device don't even have to come into play for this sort of attack or compromise. A great example is the 2008 Baku-Tbilisi-Ceyhan pipeline explosion in Turkey reported by Bloomberg in 2014 and subsequently covered by other news outlets (Figure 3.11).[20]

Figure 3.11 2008 Turkish pipeline explosion. (Copyright Reuters. Used by permission.)

Instead of receiving digital alerts from sensors placed along the line, the control room didn't learn about the blast until 40 minutes after it happened, from a security worker who saw the flames, according to a person who worked on the probe.

As investigators followed the trail of the failed alarm system, *they found the hackers' point of entry was an unexpected one: the surveillance cameras themselves.*

The cameras' communication software had vulnerabilities the hackers used to gain entry and move deep into the internal network, according to the people briefed on the matter.

Once inside, the attackers found a computer running on a Windows operating system that was in charge of the alarm-management network, and placed a malicious program on it. That gave them the ability to sneak back in whenever they wanted.

...

The central element of the attack was gaining access to the operational controls to increase the pressure without setting off alarms. Because of the line's design, the hackers could manipulate the pressure by cracking into small industrial computers at a few valve stations without having to hack the main control room.

The presence of the attackers at the site could mean the sabotage was a blended attack, using a combination of physical and digital techniques. The super-high pressure may have been enough on its own to create the explosion, according to two of the people familiar with the incident. No evidence of a physical bomb was found.

Having performed extensive reconnaissance on the computer network, the infiltrators tampered with the units used to send alerts about malfunctions and leaks back to the control room. The back-up satellite signals failed, which suggested to the investigators that the attackers used sophisticated jamming equipment, according to the people familiar with the probe.

Investigators compared the time-stamp on the infrared image of the two people with laptops to data logs that showed the computer system had been probed by an outsider. It was an exact match, according to the people familiar with the investigation.

The IoT system here, the networked video surveillance system comprised of multiple networked video cameras – each with the ability to compute and network in addition to sensing (video in this case) – had vulnerabilities and those vulnerabilities in the computing/networking aspects of the respective devices provided entry points into the system. More accurately, the system provided multiple entry points through its multiple cameras. This is the notion of attack surface, an array of attack points or entry points into a system. Those many cameras along the pipeline increased the opportunities, the vantage points, and the stepping stones to provide entry into the broader industrial control system (ICS).

So, an assessment of increased institutional attack surface subsequent to an installation of a new IoT system involves looking at the additional exposure that each individual IoT device brings and what the aggregate of device deployments brings to the institution or city for a particular IoT system.

Another Aspect of Attack Surface – Surface Porousness

While hopefully not overextending the metaphor, we can also consider attack surface porousness as an examination. For our purposes, a more porous surface is one with bigger holes.

> Each individual IoT device, that in aggregate makes this attack surface, can be thought of as a hole in the surface.

Depending on how well or how poorly the IoT device is configured corresponds to the size of the hole. A poorly configured and deployed IoT device corresponds to a large hole. A well-configured and deployed IoT device corresponds to a small hole – a device that's harder to compromise. Importantly, even a well-configured device is still considered a hole – the device has to be open enough to communicate with something for it to have operational functionality and contribute to that sought-after ROI. So, it's still a "hole," but it's smaller because it does not have additional pathways, or attack vectors,[21] into the device. Those devices with many different pathways into the device – importantly, including those pathways that do not need to be open – correspond to larger holes.

In the next chapter, we'll talk more about what makes a well-configured IoT system and what makes a well-configured IoT device.

The spectrum of device implementation quality, i.e., how well the device was implemented – configured and deployed, determines the porousness of the attack surface. The porousness is how easy or hard it is to exploit any one point on the surface.

Because the bulk of devices in IoT systems are usually installed in the same time period, often by the same people (usually vendors/contractors) using the same procedures, they are often all configured similarly. There's a good chance that if you find one or two devices poorly configured and deployed, almost all of the other devices will be deployed similarly (Figure 3.12).

To evaluate the changes to an institution's or city's attack surface then, we look at the surface itself – the number of new devices that could be maliciously repurposed – as well as the surface porousness – how big are those holes that comprise that attack surface?

Attack surface is number of attack points. Porousness is the ease or difficulty of attacking or exploiting each of the individual points that make up the surface porousness (Figure 3.13).

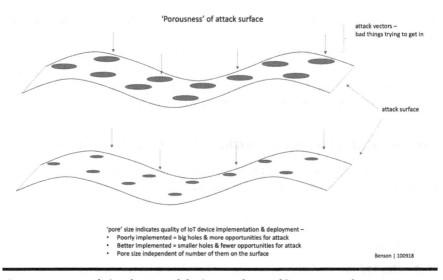

Figure 3.12   Poorly implemented devices scale – and increase surface porousness.

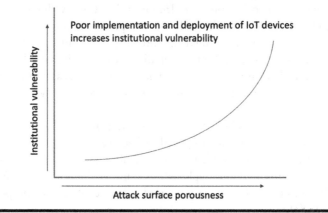

Figure 3.13   Institutional cyber vulnerability grows with attack surface porousness.

Between attack surface size – device count – and attack surface porousness – size of the holes, one is probably more in our control than the other. My sense is that we won't have much control over the number of devices deployed. When we purchase and implement an IoT system, there will need to be a sufficient number of devices, the Things in Internet of Things, to be able to accomplish the objectives for the institution or city that the IoT system purports to do. While there will be some, I don't anticipate having much opportunity to whittle that device count down because making that number of devices smaller will likely degrade what that IoT system is supposed to do.

What we can control and impact is the porousness, the size of the holes, the quality of the implementation of the Things in the IoT system. This is nontrivial and requires process and policy change, but it is within our control as institutional and city consumers of IoT systems products and services.

We can set new expectations of our IoT systems vendors and providers. We can raise the bar of what a well-configured and deployed IoT system looks like, and through contracting, hold our vendor/contractor partners to that arrangement. IoT systems vendor management and vendor relationships are discussed in more detail in Chapter 7. Determining system boundaries and acknowledging systems boundary fluidity is discussed in Chapter 5.

As we review changes to cyber risk profile of the city or institution brought about by the implementation of a new IoT system, we take into account (1) the increase in the number of devices – which corresponds to increasing the size of the attack surface, and (2) the changes to the porousness of the attack surface – the sizes of the 'holes' that each of those IoT devices create.

A useful tool for getting a quick glimpse of your city's or institution's exposure of many IoT and ICS devices is via a website called Shodan.[22] Started in 2009, the Shodan service scans all public IP addresses worldwide for ports and services typically associated with ICS systems. In recent years, the service has increased the number and types of ports and services scanned so that it captures the rapidly growing number and types of IoT devices (Figures 3.14 and 3.15).

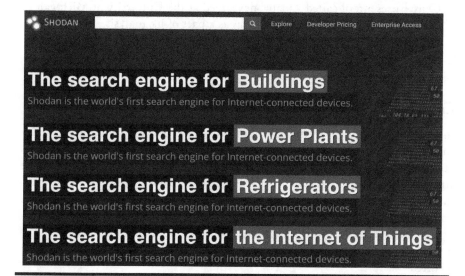

Figure 3.14 Shodan search engine searches for many types of devices.

XZERES Wind Turbine

XZERES Wind designs & manufactures wind energy systems for small wind turbine market designed for powering homes farms or businesses with clean energy.

Explore

PIPS Automated License Plate Reader

The PIPS AutoPlate Secure ALPR Access Control System catalogs all vehicles entering or exiting an access point to a site or facility.

Explore

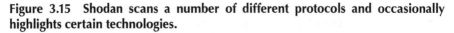

Figure 3.15 Shodan scans a number of different protocols and occasionally highlights certain technologies.

The Shodan search engine is freely available and open for anyone to use. Additional services are offered as a part of various paid services. Shodan makes some of these additional services available at no charge for academic users. In addition to access via a web user interface and command-line interface, Application Programming Interfaces are offered for a variety of programming languages to include Python, Ruby, PHP, C#, Java, Perl, and Powershell for integrating into customized applications.

Another tool, from a different provider, is called Censys.[23] Censys was developed by researchers out of the University of Michigan. It also scans all public addresses and makes results available but takes a different approach. Censys uses a tool called Zmap[24] developed at the University of Michigan and also correlates security certificates as a part of its analysis. The Censys effort has spun out of the research work and is now being commercially developed.[25]

Changes in IoT Systems Manageability

In assessing the changes to an institution's or city's cyber risk profile, we looked at changes in the five types of traditional risk – Operational, Financial, Reputation, Strategic, and Compliance/Regulatory risk. Then, we looked at changes to the attack surface of an institution or city. Finally, we'll look at changes to the institution's or city's ability to manage the newly deployed IoT system as well as the impact of managing this system on previously deployed systems.

As discussed earlier, in the course of choosing, procuring, implementing, and managing a new IoT system, some combination or all of the following costs are incurred:

- Cost of the devices themselves
- Cost of the supporting service or service
  - Licensing costs, whether SaaS or on-premise deployments
  - SaaS service or hosting costs
  - If on premise installation:
    - VM or hardware costs
    - Hosting costs
    - Data center costs
    - Potential separate support database costs.
- Costs of service deployment
- Costs of ongoing service management, e.g., SaaS, management
- Integration costs with other institutional or city information services
  - Internally or externally borne.
- Cost of device deployment
- Ongoing licensing
- Ongoing server application or service support and maintenance costs
- Device servicing and repair
  - Whether internally performed or contracted.
- Server application upgrade costs
- Training costs for system users
- Training costs for internal system support.

The most expensive items on this list are human support costs. These costs are incurred at the time of installation, but have most impact in the ongoing support and maintenance of the IoT system. The endpoints themselves, the Things, are often relatively very inexpensive. In fact, some companies may choose not to charge for the devices and instead seek their revenue from services and use of the institution's or city's data.

The recurring cost of ongoing support and maintenance, whether incurred internally or contracted externally, can be some of the most expensive aspects of the system. Paying for the internal or external work can be substantial. By the same token, *not* providing needed support and maintenance incurs the costs of increased risk to the city or institution.

Managing the IoT System

Because human involvement in installation/deployment and ongoing support and maintenance is one of the most substantial costs in an IoT system deployment, the manageability of the IoT system has great impact.

Just a few examples of aspects of IoT system implementations that must be managed include:

- IoT service or service health and updates
- IoT client application health and updates
- Any system supporting on-premise hardware (which itself must be maintained)
- Health of network segment(s) on which the IoT system is deployed.

IoT systems that assist with these aspects, i.e., make them easier to perform, lower the demand for human interaction and intervention with the system. This, in turn, lowers the most substantial cost of the IoT system implementation and operation.

Therefore, the manageability of an IoT system has great importance to the institutional or city IoT system consumer. In turn,

> because this attribute of systems manageability is so impactful to the institutional or city consumer, it becomes an opportunity of a true competitive differentiator for IoT systems product and service providers.

We'll spend more time on the details of IoT systems manageability in Chapter 6.

Suggested Reading

1. Fiore-Gartland, Brittany, and Gina Neff. Communication, mediation, and the expectations of data: Data valences across health and wellness communities. *International Journal of Communication* 9 (2015): 19.
2. Abbott-Donnelly, Ian, and Harold Lawson. *Creating, Analysing and Sustaining Smarter Cities: A Systems Perspective*. London: College Publications, 2017.
3. Definition of VALENCE. Accessed January 15, 2019. https://merriam-webster.com/dictionary/valence.
4. Truthiness - Wiktionary. Accessed January 15, 2019. https://en.wiktionary.org/wiki/truthiness.
5. Occam's Razor | Origin, Examples, & Facts. Encyclopedia Britannica. Accessed January 15, 2019. https://britannica.com/topic/Occams-razor.
6. Van Hemert, Hendrik. Interview with Hendrik Van Hemert, Pacific Northwest Regional Director - Technical Services, McKinstry, December 17, 2018.
7. Benson, Chuck. Costing IoT Systems in Cities & Institutions. *Long Tail Risk* (blog), May 12, 2017. http://longtailrisk.com/2017/05/12/costing-iot-systems-cities-institutions/.
8. Vlaar, Paul W. L., Frans A. J. Van den Bosch, and Henk W. Volberda. Coping with problems of understanding in interorganizational relationships: Using formalization as a means to make sense. *Organization Studies* 27, no. 11 (November 1, 2006): 1617–38. doi:10.1177/0170840606068338.
9. Build Free Databases. Amazon Web Services, Inc. Accessed January 17, 2019. https://aws.amazon.com/free/database/.
10. Benson, Chuck. Organizational-Spanning Characteristics of IoT Systems. *Long Tail Risk* (blog), June 7, 2016. http://longtailrisk.com/2016/06/07/organizational-spanning-characteristics-iot-systems/.

11. Koschmann, Matthew A. The communicative accomplishment of collaboration failure. *Journal of Communication* 66, no. 3 (2016): 409–32. doi:10.1111/jcom.12233.
12. Dossick, Carrie S., and Gina Neff. Organizational divisions in BIM-enabled commercial construction. *Journal of Construction Engineering and Management* 136, no. 4 (2010): 459–67. doi:10.1061/(ASCE)CO.1943-7862.0000109.
13. Three-Body Problem. In *Wikipedia*, January 8, 2019. https://en.wikipedia.org/w/index.php?title=Three-body_problem&oldid=877333039.
14. Black, Paul E., Karen Scarfone, and Murugiah Souppaya. Cyber security metrics and measures. In *Wiley Handbook of Science and Technology for Homeland Security*, edited by John G. Voeller. Hoboken, NJ: John Wiley & Sons, Inc., 2008. doi:10.1002/9780470087923.hhs440.
15. Pfleeger, Shari Lawrence, and Robert K Cunningham. Why measuring security is hard. *IEEE Security & Privacy Magazine* 8, no. 4 (2010): 46–54. doi:10.1109/MSP.2010.60.
16. Types of Business Risk. Accessed January 20, 2019. https://smallbusiness.chron.com/types-business-risk-99.html.
17. 2017 Cybercrime Case Studies | Verizon Enterprise Solutions. Verizon Enterprise. Accessed January 20, 2019. https://enterprise.verizon.com/resources/reports/data-breach-digest/.
18. IoT: Number of Connected Devices Worldwide 2012–2025. Statista. Accessed January 21, 2019. https://statista.com/statistics/471264/iot-number-of-connected-devices-worldwide/.
19. Columbus, Louis. 10 Charts That Will Challenge Your Perspective Of IoT's Growth. Forbes. Accessed January 20, 2019. https://forbes.com/sites/louiscolumbus/2018/06/06/10-charts-that-will-challenge-your-perspective-of-iots-growth/.
20. Mysterious '08 Turkey Pipeline Blast Opened New Cyberwar, December 10, 2014. https://bloomberg.com/news/articles/2014-12-10/mysterious-08-turkey-pipeline-blast-opened-new-cyberwar.
21. What Is Attack Vector? - Definition from WhatIs.Com. Accessed January 21, 2019. https://searchsecurity.techtarget.com/definition/attack-vector.
22. Shodan. Accessed January 28, 2019. https://shodan.io/.
23. Censys. Accessed January 28, 2019. https://censys.io/.
24. The ZMap Project. Accessed January 28, 2019. https://zmap.io/.
25. Censys Raises $2.6 Million Seed Round Led by GV and Greylock Partners, November 27, 2018. https://businesswire.com/news/home/20181127005029/en/Censys-Raises-2.6-Million-Seed-Led-GV.

Chapter 4

Systems of Systems and Sociotechnical Systems

As with other complex systems in nature, Internet of Things (IoT) systems are actually systems of systems. Unlike other complex systems found in nature, IoT systems include and integrate both technical and social systems. Examples of natural systems include the human body with many various subsystems such as the nervous system, the digestive system, circulatory system, lymphatic/immune system, respiratory system, and many others. All of these systems contribute to the system of the human body. These systems also have interaction, interdependencies, and inter-relationships between them. For example, the circulatory, digestive, and respiratory systems work in concert to acquire, process, and deliver energy throughout the body. Importantly, it is the interdependencies and relationships between these systems that are critical components and aspects of the in toto system of the human body.

Other examples of system types found in nature and society include economics systems, cultural systems, thermodynamic systems, organizational systems, logical systems, and others. Many of these systems have aspects that cross over into each other's system, and the exact system, then, becomes where one draws the boundary. Often it is not as important as where the boundary is drawn as it is to ensure everyone in the process understands that is where the boundary has been drawn (and for everyone in the process not to forget that that is where the boundary has been drawn).

The Bertalanffy Center for Systems Science suggest these as descriptions of a system and/or attributes of a system:[1]

- Is "whole that functions as a whole by virtue of the interaction of its parts; roughly speaking a bundle of relations" (Anatol Rapoport[2])
- Is a set of particular interactive relationships that maintains dynamically in operation of a certain whole
- Can be physical, biological, psychological, sociological, or symbolic
- Can be static, mechanical, mechanically self-regulating, or organismically interactive with the environment
- Is a set of elements standing in interaction, where its wholeness depends on the parts, and the parts depend on the whole where one and one equals two plus
- Is a complex of components that becomes an entity through the mutual interaction of its parts, from atom to cosmos
- Is the organized relationship of the parts of a whole.

Broadly, IoT systems are comprised of systems of systems of technology systems as well as systems of systems of human organizational systems such as corporate and institutional hierarchies and bureaucracies. The integration of the two is yet another instance of a system of systems.

Technology Systems within and around IoT Systems

As discussed before, the Thing in the Internet of Things, or the T in IoT, is a device that computes, is networked, and interacts with its environment in some way. An IoT *system* can be thought of as some combination of the below:

- Set or an array of these devices (sometimes called endpoint devices)
- A server or service that acts as a data aggregator from the devices and/or a device controller
- Data storage, data management, and data treatment/processing services and processes
- Data analytics services
- Data publishing – reports, dashboards, exports, data interfaces/web services, etc.

While there are other configurations, e.g., mesh networks, that we will surely see in the coming years, for the near term, most IoT systems will look something like the diagram shown in Figure 4.1. Institutions and cities will have plenty of challenges with this type of IoT system (and future meshed systems may well be more challenging).

Reading from left to right, there are multiple things/devices which then feed into an aggregation server or service. Alternatively, or in addition to an aggregation server or service, there could also be a control server or service that provides

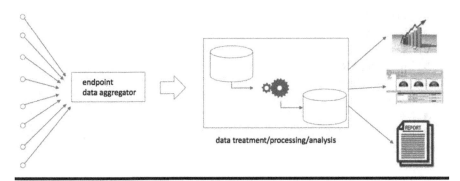

Figure 4.1 Most IoT systems currently being deployed look something like this.

commands, software updates, or others to the endpoint devices. This aggregator then provides its data stream to multistep and iterative data processes – such as data treatment/curation and management – and analysis. Finally, the processed and analyzed results are packaged and formatted for distribution and publication for sharing with combinations of specific and broad audiences.

A more specific example of an IoT system could be an energy metering system for an institution or city. In the example of a university campus, these energy management systems regularly (several times an hour) measure electrical power consumption in buildings and facilities across the campus. The number of measured points continues to grow as more buildings are measured as well as increased granularity of measurement is desired or required. There are several reasons for this increase in measured points and increased granularity of sensing of energy usage:

- The desire for increased accuracy of distributed usage (which supports increased accuracy of distributed costing and billing)
- Government ordinances and requirements
- Building certifications.

Certifications and Regulations Can Drive the Shape of IoT Systems

One example of building certification is Leadership in Environmental and Energy Design (LEED). LEED has its roots dating at least as far back as 1993[3] and is the most recognizable and widely used "green building rating system" globally though new, augmentative, or competing certifications are also available.

Other green building certifications that include energy management as a factor include (but are not limited to):

- BOMA 360[4]
- BREEAM[5]
- Class G[6]
- Energy STAR[7]
- Green Globes[8]
- Living Building Challenge[9]
- NGBS[10]
- PEER[11]
- SERF.[12]

Some LEED energy metering requirements include:

- Having in place a computer-based building automation system (BAS) that monitors and controls major building systems, including, at a minimum, heating, cooling, ventilation, and lighting
- Have a preventive maintenance program in place that ensures BAS components are tested and repaired or replaced according to the manufacturer's recommended interval
- Demonstrate that the BAS is being used to inform decisions regarding changes in building operations and energy-saving investments
- Develop a breakdown of energy use in the building … by using energy bills, spot metering, or other metering to determine the energy consumption of major mechanical systems and other end-use applications
- Based on the energy-use breakdown, employ system-level metering covering at least 40% or 80% of the total expected annual energy consumption of the building
- Permanent metering and recording are required
- Others.

The energy sensing, measuring, monitoring, and data collection requirements changes that many of these certifications, such as a LEED mandate, are quite a different scenario from the requirements of the past. For example, prior to the rapidly evolving and growing energy management systems that we see now, measuring the electricity usage of a building, campus, or city was fairly rudimentary. In most cases, cross-campus energy cost determination and analysis of usage came from measuring electricity usage at the entry point of the power source to campus and then dividing that number by the gross square feet of buildings and spaces on campus that received electricity in some way. That was as close as cities and institutions could get to a granular estimate of energy usage and it revealed very little about actual usage from different areas of the campus.

In current (and rapidly evolving) energy management systems – which are IoT systems – energy meters (usually electrical) – which number in the thousands or tens of thousands or more – are the devices. These devices are the T in this

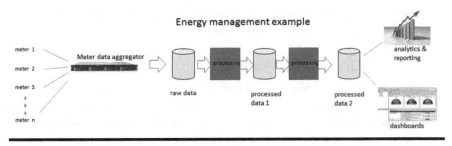

Figure 4.2   An IoT energy management *System* example.

IoT system, and *each of them is a networked computer* that also interacts with the environment in some way. In this case, the interaction with the environment is measuring electrical power – usually in at least a few different ways. Types of measurements a single meter might read include:

■ Voltage
■ Current (amperage)
■ Power (watts, watt-hour)
■ Volt-ampere reactive
■ Frequency
■ Total harmonic distortion (power quality)
■ Voltage surges and sags
■ Fault events
■ Others.

A single meter, then, can produce many sensed points which become data points. (As a side note, this fact can lead to real system impacts when administrative decisions are made, for example, to increase from three measured points to ten measured points across a large number of actual devices. The amount of data to be managed can be increased by several multiples with a single administrative or operational decision.) (Figure 4.2)

The energy data from the meters is collected by the aggregator. There are an increasing number of aggregation approaches. For example, one of the traditional products, though still evolving, is the Tridium Niagara framework.[13] Newer systems include companies such as Optio3.[14]

Advanced Metering Program at the University of Washington

At the University of Washington (UW), we established an Advanced Meter Management Program upon the conclusion of the Pacific NorthWest Smart Grid (PNWSG)[15] effort completed in 2015 of which UW was a participant. The project was the largest of 16 Smart Grid demonstration projects funded by the Department of Energy under the American Recovery and Reinvestment Act.

The effort spanned five states – Washington, Idaho, Montana, Oregon, and Wyoming – and involved 60,000 metered customers.

Operationalizing the PNWSG Project

At the end of the project, the objective has been to operationalize the effort so that initial project investment could be leveraged to optimize campus energy usage, minimize greenhouse gas, create more effective energy billing mechanisms, integrate with monitoring-based commissioning programs, develop data analysis programs, and collaborate with existing university programs such as the campus's sustainability office. Supporting activity for these objectives include:

- Troubleshooting existing energy meters
- Installing new energy meters
- Developing energy system provider criteria
- Maturing knowledge of internal systems and subsystems
- Establishing an approach and rhythm for systems support and scaling
- Integrating BASs data with energy systems data
- Developing, implementing, and integrating energy analytics, building automations systems data analytics
- Coordination with multiple interdisciplinary organizations, individuals, and institutional programs
- Developing the capacity to organizationally work with multiple evolving technologies
- Developing contract language around data ownership for external vendor relationships
- Developing organizational maturity and capacity to identify which vendors/providers to explore relationships with (and who to turn away)
- Annual development, review, and implementation of vendor contracts
- Others.

Approach

The approach includes a multidisciplinary and multi-organization team that meets every two weeks to address meter outages and performance, data management and stewardship, interdisciplinary and interorganizational, coordination, client needs, program scalability, new technologies, and others. The meeting agenda is formal and is codeveloped by university staff and the major vendor/provider.

Participants in the biweekly meetings include:

- The university's Energy Conservation Director
- The power distribution manager (electrical engineer on university staff)
- The heating ventilation and air conditioning HVAC/BAS manager (mechanical engineer on university staff)
- Facilities management IT leadership
- HVAC/energy management systems vendor/provider
- Subcontractor for HVAC/energy management systems vendor/provider.

Challenges

Early challenges to the operationalization effort included aspects as basic as meter outage troubleshooting. Early in the program, when meters failed, it was often not clear where the failure occurred.

For example, some of the types of failure included:

- Failure at end point
  - Meter failure or substantial degradation
  - Meter misconfiguration resulting in failure
  - Physical network connectivity at the point of installation, e.g., the network connection was unplugged or plugged into incorrect jack
  - Virtually, all of these types of failure require a skilled, trusted technician – whether internal or contracted – to go to the actual endpoint and investigate and subsequently initiate repair. (This takes time, requires skilled labor, and is expensive.)
- Failure en route on network path
  - Failure on network path from device to meter data aggregator.

Lessons Learned

Some of the lessons learned included:

- IoT device and data point naming conventions. Establishing and managing a plan and schema for numbering and naming all of the metered points (generated by IoT devices) is complicated and requires thoughtful planning
  - Inconsistencies in the IoT device and data point naming schema make it more difficult and costlier to support the IoT device and the data points generated from them. Lack of a standardized naming schema in a single building or across buildings makes it more difficult to scale the system.
- Aggregation. The number of IoT devices and the data points that they generate can grow very rapidly. And the data aggregation approach may soon become insufficient and/or excessively difficult to work with

■ Technologies and business models evolve rapidly, and it is possible to end up with a no longer supported technology. It is important to plan for this uncertainty
■ Developing strong, iterative, working relationships across internal (i.e., multiple departments and groups) and external (i.e., contractors/vendors) teams is critical to success and is not easy to do. It takes work.

IoT System Don't Live Alone – They Need Networks

The IoT system described above does not exist independently. Without an underlying supporting substrate of network system, the IoT system could not operate.

In the past, the phrase "network backbone" might have been used to describe the underlying network technology. However, today's networks, interconnected network segments, and interdependencies are much more complex than earlier, smaller, and more basic network hierarchies. In fact, one of the attributes of poorly designed and considered IoT systems products and services provided by some vendors is that

> they make the flawed assumption that today's networks found within cities and institutions are relatively simple, homogenous, and similar to networks from 20 years ago.

Further, many of these IoT systems "solutions" approaches that are poorly construed seemed to have been designed and perhaps tested on simple networks that do not reflect today's much more complex and evolving networks that intertwine legacy networks and network segments with newer networks and their network segments.

This aspect of an IoT systems provider approach – the understanding and awareness that the institutional consumer's networks may be complex and laced with old and new – can be a substantial differentiator for provider/vendor competitiveness in the market (Figure 4.3).

It is not uncommon for cities and institutions to have a variety of different networking equipment, technologies, and capabilities across their respective spaces.

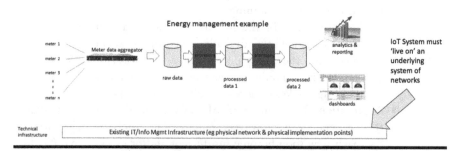

Figure 4.3   IoT systems are dependent upon underlying networks.

This fact can arise somewhat naturally as oftentimes cities and institutions do not have the funding, skill sets, or time to upgrade all of the networking equipment – and the concomitant technical and management skill sets needed – at once. Similarly, when the network grows – for example, when a new building comes on line – the desire will be to use relatively newer networking technologies with more capabilities and possibly cheaper than to try to match the networking technologies used when an existing building was outfitted for network 10 years ago. *The combination of these historical factors can result in an overall network that has uneven capabilities and uneven management capacity across the organization.*

This complexity of an institution's or city's network system is exacerbated by something on the other end of the technology time spectrum. It is the ease with which network segments can now be created and deployed. Technology that has enabled network segments such as Virtual Local Area Network[16] and virtual routing and forwarding[17] has enabled logical segmenting of networks for use with different systems.

In the past, either the technology for certain types of segmentation didn't exist or it was too labor-intensive or required such specialized skill sets that networks were hard to segment. Because they were hard to segment, there weren't as many segments that were deployed. Now, however, network segments are much easier to deploy – particularly with newer technologies such as Software-Defined Networking[18] – that there are many more network segments deployed and that rate of growth will likely increase.

Historical benefits of network segmentation include increased network performance, network access management, managed visibility of the network to other networks, and the ability to decouple segments of the network so that some failures in one network segment are not transmitted to another.

Network segmentation has been widely touted as a solution to virtually all IoT risk and security issues. However, there are at least two major issues with the growth in number of network segments:

- Network segments require management, which in turns draws on the pool of systems management resources – e.g., skilled staffing – and most organizations do not have the management capacity to fully manage all of those new network segments
- IoT systems are of such broad variety that attempting to create 1:1 ratios of IoT systems to network segments will further accelerate the network segment count and further push the systems manageability challenges.

Data Systems

Data systems are another type of technical system. There are multiple approaches and methods to store, transmit, and receive data, a myriad of data protocols and schemas, and multiple places to store data and a large variety of sizes of data stores.

For example, an IoT device might have onboard storage of a few kilobytes upwards to low gigabytes. The data on that device might be stored in one format whether a standard format or proprietary. In turn, that data will be transmitted over some interface.

Examples of storage formats are

- File systems (many options)[19]
- Microsoft SQL Server[20]
- Oracle[21]
- MySQL[22]
- Others.

Examples of cloud data storage options are

- Microsoft Azure[23]
- AWS[24]
- Google[25]
- IBM[26]
- Microsoft OneDrive[27]
- Dropbox (formerly via AWS)[28]
- iCloud[29]
- Others.

Examples of data connection protocols are

- MS SQL server connection protocol
- Oracle connection protocol
- Open database connectivity (ODBC)[30]
- Java database connectivity (JDBC).

Examples of network protocols are

- Modbus[31]
- BACnet[32]
- Fox 1911[33]
- http 80 or 443
- Others.

IoT systems have many different components, sections, and subsystems. In turn, IoT systems must work in concert with existing and planned institutional or city networking systems. These networking systems, again in turn, have many interconnecting network segments and management complexities. With technology alone, we have substantial systems of systems considerations. It starts to get really interesting, though, as we look at human and organizational systems as systems of systems.

Organizational Systems

Institutions, cities, and corporations are comprised of departments, functions, roles, centers, and hierarchies. All of these, in turn, have communication methods, histories, processes, procedures, policies, and many other attributes that have evolved over time to assist the institution, city, or corporation in meeting their respective needs. The increase in the number of these facets and the interrelationships between them increase the organizational complexity of the organization.

Also, notably, some components that originally evolved to satisfy a particular need or set of needs can have degraded or nonexistent utility when the original need goes away when the organization's objectives and/or motivations change. Unfortunately, some of these components that have organizational costs to maintain can remain long after they are needed and create unneeded baggage for the current system. That is, they have a cost, but no longer provide any offsetting value.

There is an aspect of this that is sometimes called, "the pot roast principle" or pot roast parable. A version of it is recounted in a Psychology Today article*:

> One day after school a young girl noticed that her mom was cutting off the ends of a pot roast before putting it in the oven to cook for dinner. She had seen her mom do this many times before but had never asked her why. So, this time she asked and her mom replied, I don't know why I cut the ends off, but it's what my mom always did. Why don't you ask your Grandma?
>
> ... So the young girl called her grandmother on the phone and said, Grandma why do you cut the ends off the pot roast before cooking it? Her grandmother replied, I don't know. That's just the way my mom always cooked it. Why don't you ask her?). So, undeterred, the girl called her great grandmother, who was living in a nursing home and asked her the same question - why did you cut the ends off the pot roast before cooking it? ... She said, "when I was first married we had a very small oven, **and the pot roast didn't fit in the oven unless I cut the ends off.**"

There are many examples of the pot roast principle or institutional inertia in our cities and institutions. Often there are much better, more elegant, more supportable ways of doing things and newer technologies to support them. However, because change within institutions and cities is usually difficult, these old methods

---

* https://www.psychologytoday.com/us/blog/thinking-makes-it-so/201402/
the-pot-roast-principle

stay in place. It takes real institutional energy to change them. For this reason, many old legacy approaches and technologies remain, and newer approaches and technologies become intertwined with them.

So, in addition to the many current, evolving relevant policies, processes, and approaches – cities, institutions, corporations, and other large bureaucracies, also must contend with residual artifacts that have a cost (e.g., the time taken to cut the end off of the roast) that no longer provide value.

All of these interdependent components, relevant and evolving – as well as old and no value – contribute to the creation and sustainability of systems within systems.

Organizational Systems within a Higher Education Institution

Higher Education institutions, such as universities, have a multitude of departments, divisions, and other organizational systems within systems. For brevity, a partial list of the many complex organizational structures within the administrative aspect of higher education institutions includes:

- Planning and budgeting
- Finance and treasury
- Capital development - building/upgrading/renewing buildings, additions, labs, and others
- Facilities management – operating and maintaining those buildings and spaces
- Central IT
- Distributed IT
- Police/law enforcement
- Housing
- Food services
- Senior leadership – president, provost, and others
- Trustees, regents
- Others.

In turn, each of these have further components. For example,

- Central IT may have these different divisions:
  - Networking
  - Infrastructure
  - Teaching and learning support
  - Research support

- Application support
- Cybersecurity and risk mitigation
- Many others.
■ Facilities management organizations may have these divisions:
  - Building and space maintenance
  - Building and space development/construction
  - Engineering
  - Operations
  - Custodial services
  - Transportation and parking
  - Many others.
■ Capital development may have these divisions:
  - Space management & planning
  - Project management
  - Real estate management
  - Many others.

These systems also have substantial interdependencies that contribute to a larger system for a system of systems. For example, these organizational units/structures can have the interdependencies shown in Figure 4.4.

IoT Systems are Often Sociotechnical Systems

This aspect of IoT systems being intertwined in systems of systems of both the technology type and organizational/human type qualifies them for another descriptor – *sociotechnical systems*.

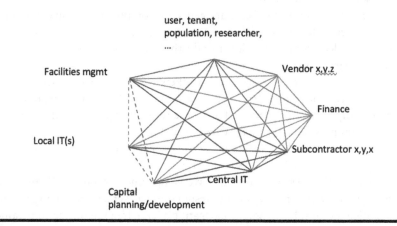

Figure 4.4 Many interactions between many organizations.

Trist and Bamforth[34] introduced the concept of sociotechnical systems and coined the term of the same, in the early 1950s in the course of analyzing work in English coal mines. From Pasmore.[35]

> The term 'sociotechnical system' was coined by Trist to describe a method of viewing organizations which emphasizes the *interrelatedness* of the functioning of the social and technological subsystems of the organizations and the *relation* of the organizations as a whole to the environment in which it operates ... the sociotechnical system perspective contends that organizations are made up of *people* that produce products or services using some *technology*, and that each affects the operation and appropriateness of the technology as well as the actions of the people who operate it.[35]

Pasmore goes on to call out this observation from the Trist and Bamforth paper:

> So close is the relationship between the various aspects that the social and psychological can be understood only in terms of the detailed engineering facts and the way the technological system as a whole behaves in the environment ...

In more recent analysis on sociotechnical systems, Kim Vicente calls out eleven (11) interrelated characteristics of complex sociotechnical systems from the book, *Cognitive Work Analysis: Toward Safe, Productive, and Healthy Computer-Based Work*[36]:

■ Large problem spaces. "... complex sociotechnical systems tend to be composed of many different elements and forces. As a result, the number of potentially relevant factors that designers and workers need to take into account can be enormous."
■ Social. These complex systems are generally composed of people who must work together for the system in toto to work properly. "This creates a strong need for clear communication to effectively coordinate the actions of the various parties involved."
■ Heterogeneous Perspectives. "... workers in a complex sociotechnical system frequently come from different backgrounds and thus represent the potentially conflicting values of a diverse set of disciplines." Differing value structures must be resolved for decision-making.
■ Distributed. Social coordination can be complicated by people being geographically distributed, even globally.
■ Dynamic. These systems are can have "long time constants." There can be substantial delay for the system to respond to worker action, forcing workers to anticipate lags. For example, navigating a large freighter in the constrained space of a harbor or port requires knowledge of inertial lag, ship responsiveness, and environmental characteristics such as water flows, tides, hazards,

and obstacles. It is for this reason that harbor pilots are used to navigate and operate in this complex environment.

■ Hazard. There can be real danger to life and limb when inappropriate human beliefs manifest themselves catastrophic results.

■ Coupling. "... complex sociotechnical systems also tend to be composed of many subsystems that are highly coupled (i.e. interacting). *This makes it very difficult to predict all of the effects of an action, or to trace all of the implications of a disturbance* because there are many, perhaps diverging, propagation paths."

■ Automation. Sociotechnical systems tend to be highly automated with algorithms controlling behavior and worker responsibility to monitor the state of the system and handle abnormal situations. Because these abnormal situations are, by intention, exceptions to the rule, it is not routine work for the worker where psychomotor skills suffice. Further since the outcome of the abnormality could be highly impactful, the worker must engage cognitively and quickly to a situation that could have catastrophic consequence and for which the worker is little practiced.

■ Uncertainty. There tends to be uncertainty in the data available to the workers. One example might be imperfect physiological sensors in an operating room. This results in the phenomena that "... the true state of the work domain is never known with perfect certainty ...."

■ Mediated interaction. "... it is often the case that the goal-relevant properties of a complex sociotechnical system cannot be directly observed by human perceptual systems unaided ... As a result, the computer interface must serve as a 'window' into the sociotechnical system providing workers with a mediating representation of the work domain...." Often, people do not have these mediated interpretation skills.

■ Disturbances. Workers must handle unanticipated events for which they are not trained. Sociotechnical systems are designed to operate, mostly, within certain parameters. And, in turn, most systems most of the time work roughly within those parameters. Because of this, the human interacting with the system is experienced, even practiced, within these bounds. When events occur outside of those parameters, the human operator is not well versed in handling these anomalous events and extra cognition is required – to include an additional amount of cognition that the operator may not have. "As a result, information system design cannot be based solely on expected or frequently encountered situations."

Vicente reiterates that not all complex sociotechnical systems are heavily rated for each of these factors. That said, all complex sociotechnical systems will rate highly on at least some of these dimensions.

IoT systems in general can be readily mapped to these characteristics of complex sociotechnical systems brought forward by Vicente. While different IoT systems will have different weighted effects of each attribute, each of these attributes or characteristics has a place of any IoT system of scale. Some examples include:

■ Large problem spaces. Without a doubt, IoT systems constitute very large problem spaces. Year over year rates of growth for IoT devices can be found to range between 20%[37] and 35% (Figure 4.5).[38] While most predictions for rates of IoT device count growth don't claim to go beyond 5 or 10 years, what if we did use these projections just to exercise the math? For example, an institution or city with 5,000 devices today could expect to see these levels of increase of device counts.

■ By the way, it is reasonable to predict that institutions such as large research universities will skew even higher because they have city-like IoT systems for convenience, safety, health, efficiency, and others and they will also have substantial IoT systems deployments supporting research grants and activities – substantial in both number of IoT systems and number of IoT devices in those systems.

	20%	25%	30%	35%
5 years	12,441	15,259	18,565	22,420
10 years	30,959	46,566	68,929	100,533
15 years	77,035	142,109	255,930	450,792
20 years	191,688	433,681	950,248	2,021,368

■ Social. This characteristic of complex sociotechnical systems, which contains a core component of required effective communication, also very much applies in IoT systems. The organizational-spanning characteristics of IoT systems discussed in Chapter 2 are a prime example. Because IoT systems span many different organizations in an institution's or city's organizational hierarchy, effective communication and mutual understanding between them are essential.

■ Heterogeneous Perspectives. This attribute definitely applies in the IoT systems context, a complex sociotechnical example. The example that immediately comes to mind is that of the historical, cultural, and experiential differences of
    – Traditional trades professionals – electricians, carpenters, plumbers, and traditional engineering professionals
    – IT professionals
    – Operational technology – the amalgamation of the first two.

■ As discussed in Chapter 6 all of these professions have very different backgrounds and perspectives, and yet, they must all come together to select, procure, deploy, and manage IoT systems.

■ Distributed. This attribute also applies to IoT systems as support staff, operators, citizens, and others will be disbursed throughout an institutional campus, city, or region (or in some case larger). This can exacerbate the communication challenge as discussed in the Social characteristic above. This geographic distribution of support people, operators, citizens, "users," and others across a campus or across a region can also introduce cultural differences which further impact communication effectiveness and sensemaking.[39]

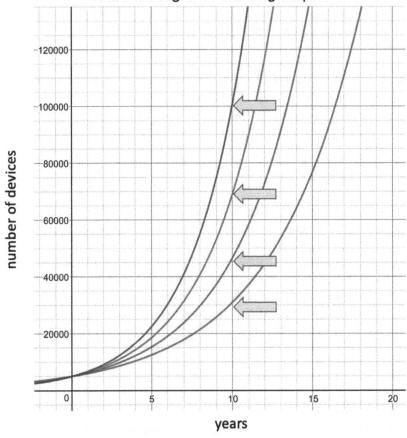

Device count growth at a range of predicted rates

Figure 4.5   Exponential growth rates at 20%, 25%, 30%, and 35% – substantially different device counts at the 10-year mark.

■ Dynamic. This awareness and management of systems lag applies to IoT systems as well, though more so at the organizational systems level than the physical/technology systems level. We can use an environmental control system for managing and monitoring a laboratory in research university as an example. One group of users/researchers may want an additional feature or feature set for a particular deployed IoT system. For example, they may want a new suite of sensors that require additional interaction with the vendor's servers and services. To do this, it must be determined that this addition does not interfere with other aspects of the system. For example, those staff involved with regulatory compliance will want to ensure that the new sensor functionality is compliant with system expectations and does not disrupt other aspects of the system already be monitored and managed for compliance. Similarly, this new set of sensors with new functionality might require modifying the supporting network segment or segments, firewalls, and possibly other aspects of the university's network infrastructure. In both cases, organizational coordination will be required, and this can take time in large, bureaucratic institutions. IoT systems users, support, and operators must anticipate this time lag.

■ Hazard. There is certainly the hazard parallel in large complex IoT systems. For example, a traffic management system in a city that uses vehicle sensing and road signaling to optimize traffic flow, both volume and speed, can have disastrous results if the system fails whether from complete failure to operate or, worse, from failure resulting in unpredictable system actions or failure resulting from compromise and operation by a malicious actor.

■ Coupling. Coupling exists between the technical components of IoT systems. Coupling also exists between the human/organizational systems that support and operate these IoT systems or that are otherwise affected by them. Finally, coupling also exists between the aforementioned technical and human/organizational systems.

One example of coupling does not directly involve IoT systems, but does demonstrate how coupling between systems can have impactful unintended consequences. In October 2013, a 16" water main pipe ruptured in a large city in a shopping area near a large university.[40] In addition to concern about electrical systems, fire sprinkler systems, and flooded parking lots keeping shoppers away, there were at least two unintended cascading effects stemming from the water pipe rupture.

Some buildings on the nearby university campus had pressure sensors on the water supply side of the building. These sensors were in place to sense water pressure drop caused by building water sprinklers activating in response to a fire. The intent was to sense water pressure drop caused by fire and activate local aural and visual alarms, as well as call the fire department.

The cascading effect from the off-campus water pipe rupture occurred because the water pressure-sensing devices on the campus buildings sensed and reacted to low pressure which had nothing to do with behavior/environment within the building, but rather due to an event that was an event that happened as a part of city systems and not even on campus. (On a related note, this is also an example of issues, signals, and events propagating between organizational systems, in this case between an institution and a city.)

The second part of the cascading effect was that the aural and visual fire alarms triggered by the sensing of low water pressure caused students, faculty, and staff to hurriedly evacuate a number of the buildings. (And then when they got outside, try to figure out why they were all outside from a multitude of different buildings.)

This substantial event – where a city water pipe ruptures quickly and causes students, faculty, and staff to evacuate buildings from a nearby campus (when there was no fire) – occurred with primarily analog systems with limited connections. Imagine possibilities where there is the digital coupling of technology over wired and wireless networks and physical systems are potentially vastly more interconnected.

▪ Automation. This aspect of complex sociotechnical systems also applies to IoT systems. By design, IoT systems are generally highly automated. In fact, that is much of the value proposition – environments are sensed, monitored, sometimes controlled by automated systems. A complex HVAC system, for example, would have very limited value if any sensed differential between desired temperature and actual temperature in a room in a building required a person to run down to the HVAC unit in the building and adjust the thermostat. Generally, IoT systems are supposed to do that kind of thing for us. (Otherwise, what was the intent of the investment?).

▪ Uncertainty. This is very much true also in the IoT systems domain. With the hundreds, thousands, tens of thousands, and more of sensors and other IoT devices and the paths that data must traverse – whether environmental data collected or actuating control data – there are many opportunities for devices to fail and paths for that data to fail.

Think of the old movies where the pilot is tapping on a cockpit instrument gauge because the reading doesn't seem to make sense. The pilot rapidly taps the gauge because they are fairly sure or fervently hoping that particular instrument is misreporting.

Any single device can fail, and many points along that data path can fail:

- Did the sensor/meter completely fail (and indicate that it did so)?
- Did the sensor/meter fail partially? Give erroneous readings? (and not indicate that it did so?)
- Did the connection to the network fail?
- Did something along the network path fail?
  - Switch, router, wiring, wireless path, authentication, and others?
  - Is the connection full off? Or intermittent? Periodic? Something else?

  − Did something fail at the point of aggregation or control?
    • Was a device not configured in the aggregating server or service?
    • Or misconfigured?
    • Was it more than one device? For example, did the same technician misconfigure a whole subset of devices?

■ There are multiple opportunities for failure along the data aggregation or control data path of any single sensor or actuator. The failure or degradation opportunities for the IoT system increases with the number of sensor or actuators that the system deploys.

■ Mediated interactions. This applies to IoT systems as well. The health of IoT systems is almost entirely monitored via mediated interactions. In the case of an energy management system where data from thousands of electrical power sensors is monitored across a campus or city, health of any single path or paths in aggregate is tracked via a mediating device or mediating system. For example, if an energy meter fails to report or reports with sufficient anomaly, often a trigger sends an alert email and/or on a computer control console of some sort.

  In the case of an alert email, some number of diagnostics may or may not be described via text in the email. The text in that email needs to be read by a human operator, interpreted by a human operator, and that human operator will need to take action on the email. The skill set and experience of that human with the multiple aspects of the system in question can greatly impact the action outcome of reading that email. A highly experienced operator might have seen that sort of alert comes through hundreds of times and know exactly what action to take. A new, less skilled, less experienced operator may take an action or actions that have little or no impact on the problem – while consuming time and institutional resources – and they may do this several times trying different options to solve the problem. These are mediated interactions. Similar actions can occur if the problem is detected as a light, beep, or other in an application on a monitoring/control console such as a PC workstation.

■ Disturbances. IoT systems do not seem to as closely correlate with this aspect as IoT systems generally don't have that human-in-the-control-loop aspect, or at least it is not as prominent. However, IoT systems do have substantial correlation with the uncertainty and coupling attributes as mentioned above.

Borrowing from Moray and Huey[41] and their work in analyzing human factors in nuclear safety, Vicente[36] uses a diagram similar to Figure 4.6 to provide a general framework of complex sociotechnical systems.

From Moray and Huey's analysis, *Human Factors Research on Nuclear Safety,*[41] on nuclear reactor operation as a sociotechnical system,

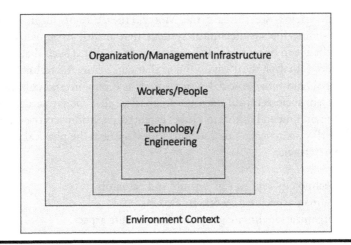

Figure 4.6   Vicente's capture of an analysis framework for complex sociotechnical systems.

> By a sociotechnical system, we mean the combination of plant hardware whose behavior is governed by physical laws, humans whose individual behaviors are governed by the laws of biology and psychology of individuals, and the interaction of the social group of the humans involved in nuclear power plant operation, management, and maintenance where the interactions are governed by the hierarchies, pressures and influences of social forces.
>
> … by a 'systems approach' we mean a way of looking at a nuclear power plant not as composed as components whose properties can be examined in isolation, but rather as a collection of components including human components, each of whose properties affects and is affected by others dynamically from moment to moment, so that to predict the performance of any component requires that one consider the state of, in general, many others.

For this analysis, the inner most system is the physical system, in this case the nuclear power plant. That system then interfaces with humans at a boundary sometimes referred to as the human–machine interface. These humans, in turn, are a part of a larger system where people in the plant, "… operate in an organizational environment that results from management decisions concerning organizational design and structure." Finally, these management infrastructures operate in a context of substantial economic, political, and social factors.

While not exactly the same, with the organizational-spanning nature of IoT technology systems, IoT systems have strong parallels to this description of sociotechnical systems and this approach can be helpful as a guideline in studying IoT systems as sociotechnical systems.

For IoT systems, we can use the Moray–Huey/Vicente diagram as a basis for understanding the sociotechnical systems that are IoT systems. A simple but important change to be made for IoT systems is to modify it such that the environment includes actual physical environment that cuts across the technology, people, organizational, and broader social systems such as economic and cultural systems. The IoT system sociotechnical diagram emphasizes that the physical environment (and other environmental factors to a lesser extent) directly impacts the other aspects.

In the IoT system version of a sociotechnical system, the physical environment directly interacts with

- The technology aspect, e.g., sensors and actuators
- The human aspect, e.g., workers, citizens
- The organizational/management infrastructure aspect.

Further, though implicit in the original, in the IoT system sociotechnical system representation, the fact that each system or supersystem is also comprised of many subsystems at that level is also called out (Figure 4.7).

For example, at the organizational/management infrastructure level, there are many subsystems/organizations that interact with each other and peer and hierarchical levels. Examples of these can be seen in Figure 4.4 earlier in this chapter.

In their respective contexts,

- Multiple technology systems can interact with each other
- Multiple people from varying backgrounds, with different skill sets, in different locations, and cultural perspectives, interact with each other

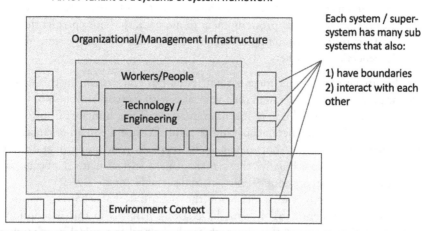

Figure 4.7   A sociotechnical framework for IoT systems.

■ Many suborganizations existing within a city, institution, or corporation interact with each other (to varying degrees of effectiveness)

■ Systems within the environmental context can interact with each other, whether physical subsystems such as weather, the local environment within a specific area of a building (e.g., a boiler room), and cultural, economic, and physical systems.

These multiple systems of systems, particularly including different types of systems intermingled with and within each other, has yet another relevant aspect included in the model – the increasingly ambiguous nature of those seams or boundaries between systems. Of particular interest is the overall systems loss that can and does occur at those seams. This systems loss directly impacts limited human, funding, and time resources, and contributes directly to the increased opportunity for lost IoT systems Return on Investment. This systems loss also contributes to degraded cyber security and cyber risk posture for an institution or city. Organizational boundaries, seams, and rich IoT systems ecosystem are discussed in Chapter 5.

Suggested Reading

1. The Nature of Systems | BCSSS. Accessed January 21, 2019. http://bcsss.org/the-center/legacy/the-nature-of-systems/.
2. Anatol Rapoport. In *Wikipedia*, December 31, 2018. https://en.wikipedia.org/w/index.php?title=Anatol_Rapoport&oldid=876117386.
3. About USGBC | USGBC. Accessed January 21, 2019. https://new.usgbc.org/about.
4. BOMA 360. Accessed January 21, 2019. https://360.boma.org/login.aspx?ReturnUrl=%2f.
5. BREEAM: The World's Leading Sustainability Assessment Method for Masterplanning Projects, Infrastructure and Buildings. BREEAM. Accessed January 21, 2019. https://breeam.com/.
6. Online Green Building & Sustainability Certification for Companies. Accessed January 21, 2019. https://class-g.org/.
7. ENERGY STAR® | Department of Energy. Accessed January 21, 2019. https://energy.gov/eere/buildings/energy-star.
8. Green Globes - Building Environmental Assessments - Welcome. Accessed January 21, 2019. http://greenglobes.com/home.asp.
9. Living Building Challenge | Living-Future.Org. International Living Future Institute (blog), October 8, 2016. https://living-future.org/lbc/.
10. National Green Building Standard (NGBS) | Home Innovation Research Labs. Accessed January 21, 2019. https://homeinnovation.com/green.
11. PEER: New Rating System for Sustainable Power Systems | U.S. Green Building Council. Accessed January 21, 2019. https://usgbc.org/articles/peer-new-rating-system-sustainable-power-systems.
12. SERF — Society of Environmentally Responsible Facilities. Accessed January 21, 2019. http://serfgreen.org/.

13. Tridium | Harness the Power of the Internet of Things with Niagara. Accessed January 21, 2019. https://tridium.com/.
14. Optio3 | Organizing the World of IoT. Accessed January 21, 2019. https://optio3.com/.
15. Pacific Northwest Smart Grid Demonstration Project. Accessed January 21, 2019. https://pnwsmartgrid.org/.
16. Catalyst 4500 Series Switch Cisco IOS Software Configuration Guide, 12.2(25)EW - Understanding and Configuring VLANs [Cisco Catalyst 4500 Series Switches]. Cisco. Accessed January 22, 2019. https://cisco.com/c/en/us/td/docs/switches/lan/catalyst4500/12-2/25ew/configuration/guide/conf/vlans.html.
17. What Is VRF: Virtual Routing and Forwarding. Plixer, December 10, 2009. https://plixer.com/blog/netflow/what-is-vrf-virtual-routing-and-forwarding/.
18. Software-Defined Networking (SDN) Definition. Open Networking Foundation. Accessed January 22, 2019. https://opennetworking.org/sdn-definition/.
19. Filesystems. Accessed January 22, 2019. https://tldp.org/LDP/sag/html/filesystems.html.
20. Microsoft Data Platform | Microsoft. Microsoft SQL Server - US (English). Accessed January 22, 2019. https://microsoft.com/en-us/sql-server/default.aspx.
21. Oracle | Integrated Cloud Applications and Platform Services. Accessed January 22, 2019. https://oracle.com/index.html.
22. MySQL. Accessed January 22, 2019. https://mysql.com/.
23. Microsoft Azure Cloud Computing Platform & Services. Accessed January 22, 2019. https://azure.microsoft.com/en-us/.
24. Amazon Web Services (AWS) - Cloud Computing Services. Amazon Web Services, Inc. Accessed January 22, 2019. https://aws.amazon.com/.
25. Cloud Storage - Online Data Storage | Cloud Storage. Google Cloud. Accessed January 22, 2019. https://cloud.google.com/storage/.
26. IBM Cloud Object Storage - Overview | IBM. Accessed January 22, 2019. https://ibm.com/cloud/object-storage.
27. Microsoft OneDrive. Accessed January 22, 2019. https://onedrive.live.com/about/en-us/.
28. Dropbox Business. Dropbox. Accessed January 22, 2019. https://dropbox.com/business/landing-t61fl?tk=sem_b_goog&_camp=&_kw=dropbox|e&_ad=278282541815|1t1|c&gclid=EAIaIQobChMImfm9kceB4AIVgsDICh3jhwU8EAAYASAAEgIuBvD_BwE.
29. ICloud.Com. Accessed January 22, 2019. https://icloud.com/.
30. What Is an ODBC Driver? Progress.com. Accessed January 22, 2019. https://progress.com/faqs/datadirect-odbc-faqs/what-is-an-odbc-driver.
31. Modbus Tutorial from Control Solutions. Accessed January 22, 2019. https://csimn.com/CSI_pages/Modbus101.html.
32. BACnet - The New Standard Protocol. Accessed January 22, 2019. http://bacnet.org/Bibliography/EC-9-97/EC-9-97.html.
33. FBI Warns Industry That Hackers Could Probe Vulnerable Connections in Building Systems. CyberScoop, December 21, 2018. https://cyberscoop.com/fox-protocol-fbi-warning-port-1911-ics-security/.
34. Trist, E. L., and K. W. Bamforth. Some social and psychological consequences of the longwall method of coal-getting: An examination of the psychological situation and defences of a work group in relation to the social structure and technological content of the work system. *Human Relations* 4, no. 1 (1951): 3–38. doi:10.1177/001872675100400101.

35. Pasmore, William, Carole Francis, Jeffrey Haldeman, and Abraham Shani. Sociotechnical systems: A North American reflection on empirical studies of the seventies. *Human Relations*, April 22, 2016. doi:10.1177/001872678203501207.

36. Vicente, Kim J. *Cognitive Work Analysis: Toward Safe, Productive, and Healthy Computer-Based Work.* Mahwah, NJ: Lawrence Erlbaum Associates, 1999.

37. IoT: Number of Connected Devices Worldwide 2012–2025. Statista. Accessed January 21, 2019. https://statista.com/statistics/471264/iot-number-of-connected-devices-worldwide/.

38. Columbus, Louis. 10 Charts That Will Challenge Your Perspective Of IoT's Growth. Forbes. Accessed January 20, 2019. https://forbes.com/sites/louiscolumbus/2018/06/06/10-charts-that-will-challenge-your-perspective-of-iots-growth/.

39. Vlaar, P. W. L., Van den Bosch, F. A. J., and Volberda, H. W. "Coping with problems of understanding in interorganizational relationships: Using formalization as a means to make sense." *Organization Studies* 27, no. 11 (2006): 1617–38. doi: 10.1177/0170840606068338.

40. http://blogs.seattletimes.com/today/2013/10/water-main-break-near-seattle-university-village/.

41. Moray, N., and Huey, B. *Human Factors Research and Nuclear Safety.* Washington, DC: National Research Council, National Academy of Sciences, 1988.

Chapter 5

Systems Seams, Boundaries, and the IoT Ecosystem

IoT systems within institutions and cities span many organizations, themselves systems, within cities and institutions, and in the course of this organizational spanning, they also necessarily span boundaries. Boundaries are roughly where interfaces occur between two or more organizational systems, two or more technical systems, or boundaries are between technical and organizational systems which gives rise to sociotechnical systems discussed in the last chapter.

Importantly, we often misunderstand organizational and technical boundaries within our institutions and cities. Often, we think and operate like these boundaries are clear, distinct, and easily delineated when, in fact, they can actually be very muddled with substantial crossover and overlap between systems.

Issues at the Boundaries

There are several aspects to boundaries – or "seams" – between systems that are important for us to keep track of and manage. Three of these include:

1. We often assume there are clear and distinct boundaries when there are not
    a. Blurring of technical networks between institution/city and IoT systems provider is one example.

2. Some organizational boundaries, roles, and functions are actually *too* distinct to plan and coordinate IoT systems implementation and management
    a. Lack of sufficient (or any) coordination between key organizations within the city or institution, e.g., central IT, facilities management, planning office, and others, is one example.
3. Systems loss occurs at these boundaries.

Forgetting that the Boundaries Are Muddled

One example of not understanding boundaries, or seams, between organizations and technical systems is the network "boundary" between the institution or city's network and that of the IoT systems provider/vendor. The traditional, and almost mathematical, view of technical boundaries between cities/institutions and their technology vendors and providers is that these boundaries provide fairly distinct and clear separation between systems – in this case, the systems constituting the institution/city network(s) and the technology vendor/provider network(s) (Figure 5.1).

The traditional, and flawed, view of network boundaries is that they are fairly distinct and well delineated. For example, in the rush of our day-to-day work lives, we can tend to implicitly tell ourselves that the network boundary is well defined and well managed, and each organization's technical system has separate, unique, distinct supporting attributes of those respective systems.

For example, we implicitly tell ourselves that the actual network of our city or institution is well defined with known and countable IP addresses, that there are well-documented and understood firewall configurations, and that there are

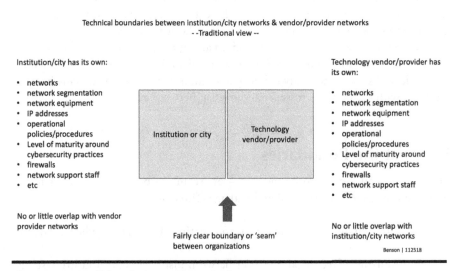

Technical boundaries between institution/city networks & vendor/provider networks
- -Traditional view --

Institution/city has its own:

- networks
- network segmentation
- network equipment
- IP addresses
- operational policies/procedures
- Level of maturity around cybersecurity practices
- firewalls
- network support staff
- etc

No or little overlap with vendor provider networks

Institution or city

Technology vendor/provider

Technology vendor/provider has its own:

- networks
- network segmentation
- network equipment
- IP addresses
- operational policies/procedures
- Level of maturity around cybersecurity practices
- firewalls
- network support staff
- etc

No or little overlap with institution/city networks

Benson | 112518

Fairly clear boundary or 'seam' between organizations

Figure 5.1 Traditional view of boundaries.

neatly controlled yet flexible and responsive firewall management processes, and current, well-documented, and well-understood cyber operational practices. Or, perhaps the institution knows that it has issues in these areas but assumes that the IoT systems provider has most or all of these things in order. Either way, there is a perception of a fairly clear boundary.

The same kind of thing can happen from the IoT systems vendor side. The technology product/service vendor or provider can have the perspective, "that's their (the client's) stuff and this is our stuff."

The problem, of course, is that this is not true. This pristine boundary is a falsehood. "There is no legitimate boundary."[1]

In her book, *Thinking In Systems,* Donella Meadows' description of boundaries in natural systems applies here as well:

> "We have to invent boundaries for clarity and sanity; and *boundaries create problems when we forget that we've artificially created them... We get attached to the boundaries our minds get accustomed to.*"[1]

This falsehood of clearly delineated and real boundaries between systems manifests itself in a number of different and impactful ways when it comes to IoT systems implementations.

In the case of technology systems, particularly network systems, the sets and subsets of networks have "boundaries" that we, in practice, assume are there but are actually not there.

Take an example case of an IoT systems vendor that provides product and services to a university for an environmental control and monitoring system in support of research. The system will have likely tens, hundreds, or thousands of networked sensors that sense and monitor a local environment through geographically disbursed sensors. The sensors could measure and report air quality conditions, temperatures of ambient air and fluids, amounts of light, vibration, and many other environmental characteristics.

The system may also have similar numbers of actuators that impact or move within the local environment in some way. The actuators might be Variable Air Volume[2] devices, damper actuators, Variable Frequency Drives[3] devices, local air heaters, remote door locks, remotely controlled video cameras, and many others.

This vendor-provided IoT system will also likely have local aggregations points, e.g., panels in a building that aggregate data from multiple sensors. The IoT system could also have on-premise support servers, e.g., Direct Digital Control[4] servers and applications. Increasingly, it is more likely that the IoT systems vendor will provide a cloud-based service, also known as Software-as-a-Service (SaaS).[5]

In this arrangement, the IoT systems vendor or provider – both technology systems and people – will "reach into" the institution's networks in multiple places, and, at the same time, the institution or city will "reach into" some of the IoT systems provider's systems – again both technology systems and people/organizational systems. For example, the IoT systems provider might query the institution's system for device health, will need to provide periodic software updates to server applications and client applications, and will hopefully provide software or firmware updates for the tens, hundreds, or thousands of deployed devices as well (though this is not yet seen in broad practice). The IoT systems provider may also make direct or indirect requests to the institution's Network Operations Center (NOC) requesting IP addresses and network configuration information such as firewall rules, network segmentation configurations and management practices, and others. The provider might also initiate trouble tickets with the institution's ticketing system (Figure 5.2).

Similarly, the institution will connect to the IoT systems providers systems and networks. They will provide sensor data for subsequent aggregation and analysis, sensors and actuators on the institution's networks might create trouble/failure alarms that are sent to the provider, and institutional support staff will contact provider support in other ways such as phone calls, emails, and text messages.

There is a lot of information sharing, and even organizational and technical infrastructure sharing, occurring.

For the institutional/city consumer of IoT systems, this problem manifests itself in cybersecurity holes, a larger and more porous attack surface, and a concomitant increased likelihood of a failed IoT system.

An institution or city is always wrestling with its cyber security capabilities and its cyber risk posture. Cyber risk posture captures a point in time of institutional cyber security capabilities, institution or city vulnerabilities, threat, attacker

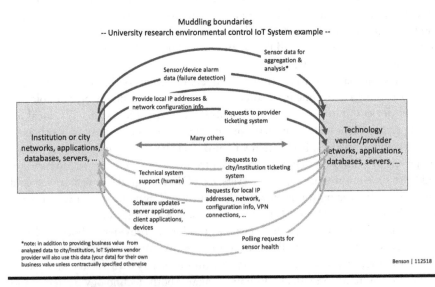

Figure 5.2 Muddled boundaries.

motivations, and other qualities. With its limited resources, an institution or city tries to enhance its cyber capabilities with efforts and programs such as

- Policy development
- User and organizational education and awareness
- Vulnerability detection
- Threat awareness/analysis
- Incident response
- Good operational practices across the institution or city
- Others.

Hopefully, the vendor/provider organization is wrestling with similar issues within their organization. *However, we frequently don't even ask how they handle these issues within their organizations!* As institutions and cities, among the time constraints, multiple projects, competing priorities, and distractions, we typically don't ask for evidence of policy, operational practices, technical controls, and similar from our vendors and providers. We just want to get the system in place, turn it on, and make it run.

Because technicians and others will have access to at least part of the city or institution's networks and systems, the city/institution should be interested in how they behave and what criteria must they meet to gain access to the networks and systems of their clients.

Intellectually, we know that the following would be reasonable things that the institution/city would want to know about the new or candidate IoT system vendor/provider partner:

- Where are your policies for cyber operations, security, and risk mitigation maintained?
  - Because those will now be applied to city/institution's network and systems.
- How do your employees learn of these policies and practices? Are they tested for awareness of and comprehension of these policies and practices?
  - Because these employees of the vendor/provider will now have access to systems and networks of the city/institution.
- Some specific policy/procedure questions could include:
  - What is the vendor/provider's password policy?
  - Are there rules for time outs for remote, e.g., virtual private network connections (VPN)?
  - Are particular certifications or levels of training and education required for those vendor staff that will be accessing networks?
  - How many different clients do the vendor/provider technicians support?
    - How do they keep them separate?

- What is the policy/procedure for a lost laptop used to connect to institution/city networks?
- What is the policy/procedure for vendor device, workstation, and systems compromise?
  - What steps are taken to ensure propagation does not reach client systems?
- What is the policy for alerting clients of device, workstation, and systems compromise?
- Many others.

The IoT systems vendor/provider's systems, policies, practices, technician training, and others will now become that of the city or institution.

> And the weakest link – whether the institution/city or provider – will have the most influence on likelihood and impact of a malicious compromise or another adverse event.

However, we more often than not don't ask these questions and, as a consequence of this omission, can suddenly have a degraded network.

The desired path for a city or institution is to develop, maintain, and enforce a robust set of criteria. However, this approach takes time and resources to implement. As such, the most important thing is to be aware of where there are issues, flaws, and weaknesses in the connections and working operation with the IoT systems vendor as the city/institution moves forward with maturing its IoT systems vendor relationship capabilities. The most substantial problem occurs – whether as institutional or city consumer of IoT systems or provider/vendor of IoT systems – *when we forget that the boundaries are muddled.* Muddled boundaries in themselves are not bad; it is when we forget that or don't take it into account that it is problematic. And, currently, institutional or city lack of awareness of muddled boundaries is more the rule than the exception.

Similarly, the IoT systems vendor or provider will want to have some idea of what they are connecting to and what the relationship will be like. Without this knowledge, the IoT systems vendor/provider

- Can expose their internal systems and networks to problematic client network and system environments
- Can propagate issues, malware, and others from one client to another
  - Creating substantial operational, reputational, and financial risk for all parties.
- Inhibit the ability/decrease the likelihood of a successful IoT systems implementation
  - Which also lends to reputational and financial risk.
- Degrade the provider's ability to help the client (institution or city) scale the system.

As with the IoT systems institutional or city consumer, the largest risk is when the vendor/provider is not aware of these issues. It is far better to be aware of these issues and risks, work to repair them, and enhance that boundary activity with a long-term plan, than to be unaware that the issues are present.

Boundary Spanners Needed – Some Boundaries Need to Be More Muddled

Another view of boundaries, this view involving different types of systems, involves those between organizations/departments where they have boundaries between each of the participating organizations and have well-defined purposes, objectives, missions, constituents, finances, operating procedures, policies, etc (Figure 5.3).

In the case of cities and institutions, IoT systems span many different organizations and departments. To meet the objectives of IoT systems implementation success (see Chapter 3), there needs to be communication, coordination, and accountability between all of the organizations. Hard seams in these cases of IoT systems are problematic (Figure 5.4).

IoT systems span multiple organizations within an institution or city in the course of the System's life cycle that includes:

- IoT system selection
- IoT system procurement
- IoT system implementation
- IoT system management and operation
- IoT system retirement.

For example, the planning and budgeting organization might be involved in designating and initially costing a system. Similarly, the finance organization may

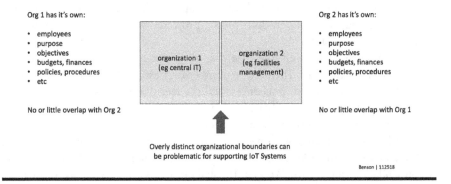

Figure 5.3 Overly distinct organizational boundaries.

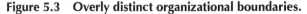

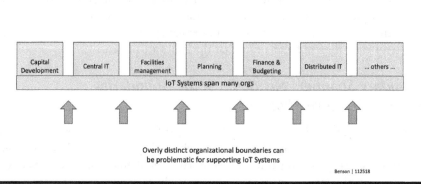

Figure 5.4 **Many overly distinct boundaries challenging for IoT systems.**

be involved in communicating funds available, potential funding approaches, and possibly procurement. If it is a large IoT system going into a new building or retrofit building, various capital development groups may be involved. During the construction, central IT may be involved regarding networking and other technical requirements. Upon completion of the building and subsequent tenant occupation of the building, the facilities maintenance organization as well as, possibly, local IT support groups will be involved to assist with running the application(s) that support the IoT system. And central IT, local IT, and facilities may be involved during the retirement of IoT system.

Ideally, almost all of these organizations would be in contact with each other at appropriate points throughout the life cycle of the IoT system and the life cycle of the building. For example, it would be very appropriate, helpful, and would serve to increase the opportunity for strong ROI and non-degraded cybersecurity posture if the facilities management organization (as one example) was able to provide input during the IoT system selection process. In this case of facilities management, the facilities professionals that would support and service sensors and actuators as well as, possibly, some of their related systems could provide feedback on effort levels to support the IoT system, risk factors during the management/operations phase, projected costs of and availability of support, and other factors.

Unfortunately, it is rarely the case that all of the organizations interact with each other along the way. There are a number of reasons for this including precedent, existing policies and procedures, and the fact that IoT systems are still new and most organizations do not have familiarity working with them, much less coordinating with other organizations on the topic.

Another core challenge with meeting, collaborating, and communicating with multiple other organizations on the topic of these complex systems is that, quite simply, collaborating is hard. And while collaboration is often touted as the de facto approach to complex efforts in complex organizations, we find ourselves implicitly or explicitly dealing with "the discrepancy between the promise of collaboration and the reality of persistent failure."[6]

In the example of civil society collaboration, Matthew Koschmann defines civil society as encompassing many interorganizational relationships among governments, nonprofit organizations and NGOs, private businesses, and local community and citizens. These types of organizations certainly apply to cities, but I think also to institutions such as universities – with their many academic units, administrative departments, centers, research facilities, and others.

Per Koschmann,[6]

> Collaboration is a hallmark of contemporary organizing, especially in the civil society sector where there is widespread recognition that the complexity and interdependence of many social issues necessitates some type of collaborative response among relevant stakeholders … [Collaboration as an area of research] *magnifies issues of trust, identity, power relations, network configurations, boundary spanning, agency and authority, negotiation*, and other key aspects of human interaction … Unsuccessful collaboration is ubiquitous in practice …

Koschmann's points are also relevant in the institution or city implementing a large IoT system. Multiple organizations or departments are involved, and collaboration and information sharing between them at various points in the IoT system life cycle are highly desirable. However, in practice, this is difficult and resource consuming, and there is a lack of shared language and little to no precedent for IoT systems selection, procurement, deployment, management, and retirement processes (much less documentation of these processes) for the city or institution. Koschmann continues,

> With an emphasis on participation, interdependence, representation, cooperation, nonhierarchical relationships, and mutual accountability, collaboration is seen as an ideal from of civil society organizing … However, in practice, civil society collaboration is incredibly complicated and frequently ineffective. Civil society *collaborations often produce limited results, involve members with contrasting goals and motivations, that are difficult to manage, are prone to gridlock and fragmentation, often do not produce intended outcomes, and can even exacerbate the problems they are trying to solve* … some collaborations fail to generate any collective action at all.

Similar views from other researchers speak even more to the indeterminate nature of knowledge within firms, cities, or institutions and the corresponding uncertainty that exists within them and the organizations between them. Per Tsoukas,[7]

> …no single agent can fully specify in advance what kind of practical knowledge is going to be relevant, when and where. Firms, therefore, are *distributed knowledge systems* in a strong sense: they are decentered

systems, lacking an overseeing 'mind'. *The knowledge they need to draw upon is inherently indeterminate and continually emerging*: it is not self-contained.

(p. 11)

Referencing R. Stacey, Tsoukas expands on the concept,

> ... the firm is a *distributed* knowledge system ... Firms are faced with radical uncertainty: they do not, they cannot, know what they need to know ... firms are not only distributed, but decentred systems – they lack the cognitive equivalent of a 'control room'[8]

This idea of a distributed knowledge system is consistent with the multiple organizations within an institution or city that must select, implement and integrate, manage/operate, and risk mitigate IoT systems (if the implementation is to be successful). In a similar vein, Bechky[9] refers to these different areas of organizational expertise as "occupational communities."

Bechky also points out how differences in language and context among occupational communities greatly contribute to the challenges of sharing knowledge across these organizations. Specifically, she points out that these three factors in conjunction contribute to the difficulty:

- Differences in language
- Context of where the organization's/occupational community's work occurs
- Conceptualization of the issues at hand (a product in the case of her research).

These factors can also be found in and between the organizations comprising the cities and institutions that are deploying and supporting IoT systems. For example,

- Differences in language can exist between a facilities organization and a central IT organizations. For example, the word "network" is ubiquitous within an IT group but rarely used within a facilities management organizations. Similarly, the word "address," when used by an IT professional, may be an IP address or MAC address. To the facilities management professional, an address might more often refer to a physical building address.
- Context of where and how work occurs applies here as well. Even seemingly more closely related organizations, or "occupational communities," such as the capital planning group and facilities group within an institution or city, can have challenges around concepts that stem from where the work is performed and the abstraction and granularity that group is accustomed to.
- For example, the concept of an "asset" to the capital planning group probably refers to a building or facility in the city or campus. On the other hand,

to the facilities management professional, the "asset" concept may well be a component of a building such a piece of HVAC equipment, a window, or a toilet. There may be hundreds, thousands, or more assets (facilities management view of assets) within a single asset (capital planning view of asset).

■ Contextualization issues can occur around how data is shared between different organizations and occupational communities. For example, a planning group may generate many computer-assisted drawings and be very comfortable working with them. On the other hand, a facilities management tradesperson may well be much more comfortable and expedient with finding a room number associated with a building number that is embedded in a work order than she/he has been given to complete a maintenance function.

Bechky continues (p. 313):

Conceptualizing knowledge in organizations with the impoverished metaphor of knowledge transfer has several implications. Simple knowledge transfer assumes a referential theory of meaning and implies that within organizations, meaning is universal and context is relatively homogenous. *Yet in practice, these assumptions do not hold.* Even when knowledge is made explicit in a codified routine, when it is communicated across group boundaries, some organizational members may not understand it *because they apply and interpret this knowledge within different contexts.* ... there is an array of meanings in organizations: *Understanding is situational, cultural, and contextual.* The creation and enactment of organizational knowledge is therefore a complex process involving the understanding of multiple communities.

Interdisciplinary Domains

Ben-Menahem[10] et al. discusses some of these boundary and multi-organizational coordination efforts in terms of interdisciplinary knowledge domains and "knowledge creation." The many organizations within a city or institution can be considered separate knowledge domains. For example, central IT organizations are experts in networks, systems, applications, and enterprise hardware. Facilities management organizations are experts in operating and maintaining the built environment from office buildings, spaces, and complex research facilities in the case of a higher education research institution. Planning groups might be experts at space planning, construction planning, budgeting activities, and others.

Further, the notion of "knowledge creation" also applies to IoT systems deployments within cities and institutions as organizations within, individually and collectively, identify institutional roles for deploying and managing an increasing number of IoT systems,

> ... multidisciplinary teams are critical for knowledge creation in increasingly specialized work environments ...[10]

From the study research on early stage drug discovery, the authors draw the conclusion that "when the interdependencies among knowledge domains are dynamic and unpredictable, specialists design self-managed (sub-)teams around collectively held assumptions about interdependencies based on incomplete information (conjectural interdependencies)."[10]

There are parallels in this research on early stage drug discovery team coordination behavior and the behavior of organizations and departments within institutions and cities as they grapple with implementing, maintaining, and operating increasingly complex and interdependent, both technically and organizationally, IoT systems.

Coordination between Organizations

Another aspect of integrating the efforts of multiple organizations within a city or institution to deploy and support IoT systems is that of coordination between and across those *organizations. Heath and Staudenmayer11 argue that, in attempts to coordinate, organizations focus more on dividing various tasks into components and assigning divisions of labor than they do on integrating those components back into a cohesive whole.* A key aspect of this failure of integration is lack of ongoing communication compounded by insufficient translation as specialists from different groups, similar to the notion of occupational communities above, have difficulty in communicating with one another.

City and institutional departments and organizations see this as well when they seek to plan, deploy, and operate IoT systems. For example, it is not uncommon for institutional planners of an IoT system to assume that central IT will manage the IoT system that sits on the networks managed by central IT when the case may be that central IT does not even know about the system! Similarly, the facilities management organization may assume and expect the capital development group will collect all of the needed building, construction, and other data, e.g., Building Information Modeling or BIM data[12] and deliver it in a consumable and actionable form, when the facilities group has not articulated much less communicated the need, the format, and the deliverable.

Heath and Staudenmeyer refer to these kinds of issues as "coordination neglect."[11]

> In order to accomplish their work, organizations must solve two problems: motivating individuals so that their goals are aligned (the agency problem) and organizing individuals so that their actions are aligned (the coordination problem) ...
>
> ... when individuals design organizational processes or when they participate in them, they frequently fail to understand that coordination

is important and they fail to take steps to minimize the difficulty of coordination ... individuals exhibit *coordination neglect.*[11]

(p. 154, 155)

Coordination neglect also reveals itself as, "People focus on the division of labor rather than on the equally important process of integration, and when they try to intervene in an ongoing process of coordination, they tend to focus on specialized components of the process *rather than attending to the interrelationships as a whole.*" (p. 171)[11]

Institutions and cities certainly see these patterns in selection, deployment, and operation of IoT systems. Each different functional group tends to focus on their own specialties and then throw systems and integration problems "over the fence" to another organization and that the integration fixes itself. Possibly worse is when an organization doesn't interact with another organization at all, such as throwing over the fence, but instead just waits for someone to fix the problem.

There are some positive efforts in aspects of this however. For example, the University of Washington has stood up a "Transition to Occupancy (T2O)" program[13] to address the transition from the near completion and completion of a new building or remodel to the occupancy and operation of that building. Among other things, this program seeks to coordinate activities between the construction and development group and the facilities operations group. It is important work. And it is also a heavy lift.

Finally, Haas[14] uses the term "organizational frontiers" to discuss boundaries where knowledge transfer occurs and how they can be managed. He describes three types of roles that can be identified in this regard: knowledge brokers, gatekeepers, and boundary spanners. Hass defines them as

■ Boundary spanner – many definitions but generally interface between internal or external areas. Facilitates intergroup/interorganizational exchanges and access to resources
■ Gatekeeper – similar to boundary spanner but might control, mediate, or guard access to information
■ Knowledge broker – participates in multiple communities and facilitates knowledge transfer.

Regarding IoT systems deployments, my perspective is that the concept of boundary spanner is the most useful within an institution or city. While there may be gatekeeper and knowledge broker aspects intrinsic, I find that the generalized concept of boundary spanner is most helpful.

Institutions and cities need boundary spanner roles. I believe it will be fairly uncommon to have dedicated, full time equivalent (FTE), "boundary spanner" positions. However, having this role be a component of existing positions will be more readily seen. The reasons for this are both financial and that boundary spanner roles embedded within existing roles support validity of the role, adding "street cred" to the role and function.

Again, these issues and observations also apply to the very environments – cities and institutions – where substantial IoT systems are being deployed. Further, some particular aspects of IoT systems make this process even more challenging:

■ Large scale of technology deployment, e.g., thousands, millions, or more of networked computing devices (sensors and actuators) and very large amounts of data to be aggregated, analyzed, and made available
■ High variability and variation of types of systems and devices and the subcomponents that comprise those devices
■ Lack of language to discuss, plan, and mitigate issues around rapidly evolving IoT systems
■ Organizational-spanning aspects of these systems
■ Novelty and embedded nature of devices and systems are very new concept and not broadly perceived or consumed
■ There is very limited precedence for successful implementation of these systems. Further, the infrastructure nature of these deployed systems can take years to discern failure.

With these challenges and impediments to coordinating across organizational boundaries and the rapidly evolving nature and complexity of these IoT systems, it is no surprise that substantial gaps exist in the selection, implementation, and operation and management of these IoT systems. There is no natural or obvious impetus for these organizational systems to independently self-organize and select and operate these systems well.

Information Loss at the Seams

These muddled boundaries, or seams, with their coordination and collaboration challenges are where substantial loss can occur within the overall system – in this case, the overall system is an institution or city. This does not mean we seek to eliminate the boundary or seam or organization, but we do need to know that this seam is there and that there is loss occurring there. With this awareness, we can seek to mitigate and reduce that loss. This loss subtracts from already finite, constrained resources and contributes to degradation in systems manageability and hence to diminished or lost ROI and increased cybersecurity issues and cyber risk exposure.

The worst-case scenario is for this loss to be occurring, growing daily, and for cities and institutions to not be aware that it is occurring as their systems become increasingly unmanageable.

This loss at the seams comes from a number of sources:

- ■ Information loss
  - – Lack of shared context
  - – Poor articulation
  - – Failure of "sensemaking."[15]
    - • Misunderstandings.
- ■ Staff time loss
  - – Excessive meetings
  - – Redundant meetings
  - – Ticket thrash
  - – Failure to ensure agreements.
  - – Opportunity cost – other productive work that staff could be performing.
- ■ Trust issues
  - – Miscommunication between parties.

These boundary and seam coordination issues, that include lost time through repetitive limited value meetings, misunderstanding, lost opportunity for trust development, disruption impact, excess "ticket thrash," and other factors, are problematic even in the case of two organizations with one boundary or seam (Figure 5.5). The bigger problem for the city or institution is that these problems quickly aggregate as cities and institutions deploy and seek to manage a rapidly increasing number of IoT systems.

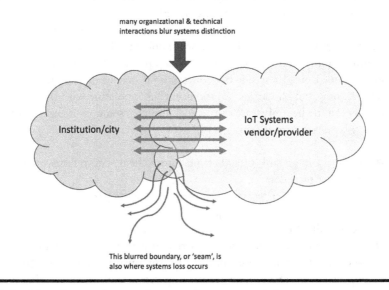

Figure 5.5   **Many interactions and transactions occur at the seam.**

For example, "ticket thrash" can be a time and resource sink and a project or process disruptor for the requestor and other parties. The idealized form of submitting a request is when help desk requests or tickets are sent to the city/institution's ticketing system or the vendor's ticketing system.[16] For very simple requests or problems to troubleshoot, the help desk can receive the ticket, take a corrective action, resolve the issue, and inform the requestor. At this point in the idealized form, the work exchange is done.

Resource demands begin to arise when the request is not so simple. For example, if the problem is not immediately solvable, research or diagnostic time may be required by the help desk. The help desk may also ask for more information. There may be a number of iterations between help desk and requestor. And during each of these iterations, typically there is some delay at each end between when the reply ticket is sent, some action is taken (whether or not the reply is immediately helpful), and the response email is provided. The more of these iterations and the time they take, generally the more disruptive and possibly intensive the request is.

The "ticket thrash" kicks into high gear, though, when three or more parties get involved, whether within the same ticketing system or across multiple ticketing systems. For example, an energy meter subcontractor may be working coordinating a network issue with the facilities client at an institution or city. In the course of doing so, the subcontractor might generate a ticket to the client's network support organization and notify (for example, via email cc) the primary contractor and the client facilities support organization. The primary contractor and the client facilities support organization may reply with affirmation, question, and/or additional information. Sometimes, response information from two or more different parties will conflict or be inconsistent. This will likely generate additional queries over the ticket from one or more parties on the ticket.

And this whole process in the preceding paragraph may repeat a number of times. Further, along the way and in the interest of hoping to clarify aspects of the issue, side bar conversations "off ticket" via separate email, phone call, or in person may develop. While in some cases, these side bar conversations may help, often they can have the effect that, now, not all parties "on the ticket" are operating with the same information. This can further obfuscate the issue of finding a solution to the original request.

So, this "ticket thrash" can

- Become a large sink of staff time across client, contractors, subcontractors, and other providers
- Be disruptive to projects and processes. If these particular requests come up at critical junctures in a project or process, they can become choke points for the overall effort
- As the number of iterations, particularly in the form of ticket messages, increases, there is more opportunity for misunderstanding. That is, more opportunities where there are varying contexts of understanding and where "sensemaking" needs to occur.

All of these aspects create systems loss. They drain on the supply of already constrained resources of the institution or city. As these kinds of issues scale across an institution or city, the loss quietly becomes substantial.

With the increased vendor count of IoT (and other) systems, there is more opportunity that many parties from many organizations will be involved in help requests, troubleshooting, and coordination attempts. These all demand more staffing resources, and increase the opportunity for misunderstanding and likelihood of disruption.

Some of this reminds one of the three-body problem[17] where predictability around two objects is generally accessible, but adding one more for three objects becomes incalculable.

The above example has been about only one request and ostensibly one ticket. IoT systems typically don't have only one issue or problem that needs diagnosing and troubleshooting. For an institution or city, there will be many IoT systems and that number will probably rapidly grow. So, picture the above in aggregate. When there are many requests across many systems, the loss is substantially higher. *Exacerbating the issue is that loss is likely not realized in the time that it is occurring* (Figure 5.6).

This look at transactions between organizations and the overall systems losses that occur and that are a natural by-product of the structure is not dissimilar to the reflections of Ronald Coase over 80 years earlier in his paper, *The Nature of the Firm.*[18] In his paper, Coase describes the balances of the costs of providing a service internal to an organization with the cost of contracting that out. Some of these

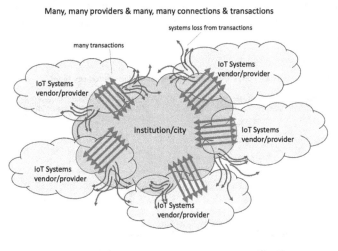

Many, many providers & many, many connections & transactions

systems loss from transactions

many transactions

IoT Systems vendor/provider

IoT Systems vendor/provider

Institution/city

IoT Systems vendor/provider

IoT Systems vendor/provider

IoT Systems vendor/provider

Institutions & cities have many – and will many have more – IoT Systems vendors and providers

-- provider relationships have many transactions – technical & human

-- transactions have costs, e.g. staff time, meeting time, disruption, opportunities for misunderstanding, trust issues, data alignment needs, technical connections to manage, security issues, etc

-- many IoT Systems means many vendors means more transactions means more loss/costs in an environment of constrained resources

-- more manageable IoT Systems require

Benson | 112518

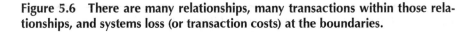

Figure 5.6 There are many relationships, many transactions within those relationships, and systems loss (or transaction costs) at the boundaries.

costs associated with these transactions include search and information costs (what is needed and with which firm to engage), negotiating costs, enforcement costs, and others. While in this book, we have been referring to this as institutional systems loss from transactions, Coase describes something fairly similar as transaction costs.

Oliver Williamson[19] substantially extended upon Coase's work, developing a field called transaction cost economics, and received the Nobel prize in 2009[20] for his work in organizational boundaries along with Elinor Ostrom[21] for her work in shared-resource systems.

Impact of IoT Systems Manageability on Loss at Organizational Seams

Because of the impacts of resource demands – staff time, technology investment, time to effect change – against a backdrop of very finite resources – e.g., limited staffing, limited funding, and limited time, the manageability of IoT systems becomes a critical aspect of IoT systems implementation success – both in terms of ROI and cybersecurity/cyber risk degradation (Figure 5.7).

IoT systems with higher degrees of manageability yield substantial value because of the enhanced ability to reduce systems loss in an environment of already confined resources. Systems with higher manageability:

■ Reduce organizational systems losses such as ticket thrash and follow-up coordinating and status-checking phone calls, emails, and texts by producing and delivering a service and product that generates fewer trouble areas, particularly those areas that require a number of different organizations to be involved within the client institution to resolve.

Impacts of IoT Systems Manageability on resource demands at vendor/institution-city seam

Issues at boundaries or 'seams'	IoT System with poor manageability	IoT System with better manageability
Ticket thrash	Uncontrolled	Reduced/managed
# of follow up & coordination calls	Many	Fewer
Opportunities for IoT System misuse (i.e. user other than intended)	High	Low
IoT System performance	Degraded, perhaps substantially	Meets or exceeds expectations
Auditability	Uncertain, questionable	Increased auditability, verifiable
Training on systems use	Difficult to use	Easier to train
# of post-deployment 'workarounds'	High	Reduced
Compliance & regulatory problems	High, poor defensibility	Reduced
Trust between vendor & city/institution	Diminished	Maintained or enhanced

Benson | 120518

Figure 5.7 Effects on IoT systems manageability at the seams or boundaries.

- Reduce opportunities for misuse by thorough and thoughtful deployment due to experience-based design, thorough and consumable documentation and training, and others.
- Have tools and support services capability and capacity available that allow the institutional or city client to evaluate the system's performance. This helps the client institution or city study and communicate ROI to constituents and funding sources.
- Have diagnostics and reporting tools that allow the city or institution to readily audit the performance and deliverable of the IoT system. This helps with both ROI communication and cyber risk mitigation for the institution. This is of particular value to IoT systems deployed and managed in regulatory and compliance settings.
- Have strong experience-based system design and deployments by the IoT systems provider/vendor, well-articulated and contractually documented expectations and deliverables by the consuming city or institution, and both vendor-based and institution/city-based service and support capacity to reduce the number of problematic work-arounds.
- Have delivered IoT systems that come reasonably close to delivering what was expected and support services capability and capacity to quickly address any shortcomings that delta between expectations and actual outcomes. These delivered systems do not degrade trust and, in fact, enhance trust in the vendor/provider relationship. Enhanced trust can lend itself to even more productive and cost-effective future interactions between the city/institution and the IoT systems vendor/provider.

IoT Ecosystems

As we have seen, the social and organizational set of interactions between even a single relationship can be complex and resource intensive over the course of the life cycle of the IoT system. However, similar to the exponential growth in the number of devices in a city or institution, there will also be ongoing substantial growth in the number of IoT systems that the city or institution must manage. And those many systems have many transactions and opportunities for systems loss, both organizational and technical. Further, many of these systems will have interdependencies between them such as couplings of the applications outputs and inputs across different systems from different providers, infrastructure components that support the IoT systems such as networks and network segments, data enmeshment, and others.

Stepping even further back, these IoT systems providers and vendors are also themselves customers of other components, products, and services – which are also product/service consumers of others. The network sharing, data sharing, organizational support sharing, device component sharing and others in this vast, growing, and multinational IoT Ecosystem, coupled with constrained resource environments, incomplete information, and intersections of altruistic and sometimes less altruistic intent, create opportunities for unexpected behavior and unexpected paths and vectors between organizations that may not even know about each other. This is a major development in our new and evolving sociotechnical world.

Malicious Actors – Another Part of the IoT Ecosystem

There is another part of the IoT ecosystem and another set of muddled boundaries and seams that we have not yet spoken of and that is the institution's or city's boundaries and seams with the vast array of malicious actors. The malicious actors could be from a large number of nation-states, sophisticated organized crime, activist groups, and others. Our institutions and cities have boundaries with them also, and they are *very* muddled. In fact, in the cybersecurity world, there is a mature and appropriate approach and mind-set that is becoming more popular and that is to *"assume breach."*[22] What this means is to:

> Assume that malicious actors are already inside your networks and are performing malicious activities with your resources, your data, and other.

By having this assumption (or being aware of this fact), the city or institution can plan proactive and reactive approaches appropriately. The traditional approach of assuming that the malicious actor is not already inside the network is problematic because it encourages institutions and cities to overinvest in fences, firewalls, figurative moats, etc. While these are an important of a cyber security approach, investments in these areas should have limits because they don't help with malicious actors that are already inside the networks.

F5 Labs[23] publishes a series on IoT risk and security called "The Hunt for IoT."[24] Some data from a recent publication, October 2018, is given as

- From data collected January 1 through June 30 2018:
    - IoT devices became the top attack target across the Internet, surpassing databases, application servers, and others
    - Spain was the top attacked country of IoT devices
    - Brazil was the top attacking country, followed by China, Japan, Poland, and the United States.
- In September 2018, digital displays at UK's Bristol airport were attacked with a ransomware attack[25], forcing the airport to communicate with travelers with handwritten schedules.
    - The displays had 39 threat actors actively connected.
- 13 IoT device-driven botnets were discovered in the first 6 months of 2018 – only six were discovered in *all* of 2017. (Botnets[26] are large networks of compromised computers, usually unknown to the computer owner, and the network is generally controlled by one source. Botnet owners/controllers can

activate thousands or millions of computers to perform criminal or nation-state hostile activity in moments.)
  – Some examples of the named IoT-driven botnets are given as (F5 Labs calls these "thingbots")
    • Wicked – targets small office/home office (SOHO) routers, closed circuit television, and digital video recorders
    • Roaming Mantis – attacks Wi-Fi routers and Android and iOS phones
    • DoubleDoor, JenX, and many others attack SOHO routers
    • Hide'n Seek – attacks IP cameras.
■ 2018 saw a rise in IoT devices use in distributed denial-of-service (DDOS[27]) attacks. (DDOS attacks disrupt normal network traffic by overwhelming the target server(s) with requests for interaction.)
■ F5 Labs and data partner Loryka[28] tested an array of IoT cellular gateways and found that 62% were vulnerable to remote attack because of weak vendor default credentials. Weak vendor default credentials mean devices/systems were installed without changing the default credentials established by the vendor. There are many, many sites on the Internet that provide freely accessible, large lists of default vendor logins and passwords.

F5 Labs goes on to report that the destructive Mirai malware – designed specifically to attack and subsequently use IoT devices in further attacks – still has a strong global presence. The F5 Labs report provides three reasons for the ongoing Mirai presence:

1. Many of the attacked IoT devices have no way of implementing firmware updates
2. The infected IoT devices are often owned/operated by people without the technical skills to appropriately configure them
3. Many of the infected IoT devices are on telecom networks and telecoms are not financially motivated to disconnect them because that would also disconnect regular paying services.

Some Action Items for Institution/City and Provider Boundaries and Seams

1. For the city/institution:
   a. How many vendor representatives, e.g., technicians and help desk support, currently have access to any one IoT system in your city or institution?
      i. Is there a list? When was it last reviewed?
   b. How many vendor representatives have access to the city/institution's IoT systems in aggregate? Tens? Hundreds? Thousands?
      i. Is there a current, managed list?

    c. How many access points – VPN's, systems, devices, network segments, etc. are open to external providers?

       i. Is this list current (e.g., at least quarterly)?

2. For IoT systems vendor/provider:

    a. How many access points does the city or institution have into your network?

    b. Are there system diagnostics available as a part your IoT system product or service that helps to reduce ticket thrash, excessive follow up calls and emails, and work-arounds?

       i. Do you have a way of tracking work-arounds?

       ii. For successive IoT systems deployments across multiple clients, is there a trend of fewer workarounds?

3. For the city/institution:

    a. Do all of the organizations that they are involved with implementing and supporting the IoT systems know that they are or will be involved?

4. For the IoT systems vendor/provider:

    a. Can you provide services and tools that can help clients measure their actual costs, internal as well as external?

       i. Being able to assist clients with helping to measure actual costs will be a differentiator amongst vendors in the highly competitive IoT systems ecosystem and marketplace.

5. For the institution/city and vendor/provider:

    a. Meet at least annually to review the IT, cyber risk, and cybersecurity practices of each other to ensure mutual understanding.

       i. The combined network is only as strong as the weakest participant.

Suggested Reading

1. Meadows, Donella H., and Diana Wright. *Thinking in Systems: A Primer*. White River Junction, VT: Chelsea Green Pub, 2008.
2. https://en.wikipedia.org/wiki/Variable_air_volume.
3. https://en.wikipedia.org/wiki/Variable-frequency_drive.
4. https://link.springer.com/chapter/10.1007/978-1-4615-4921-5_1.
5. https://searchcloudcomputing.techtarget.com/definition/Software-as-a-Service.
6. Koschmann, Matthew A. The communicative accomplishment of collaboration failure. *Journal of Communication* 66, no. 3 (2016): 409–32. doi:10.1111/jcom.12233.
7. Tsoukas, Haridimos. The firm as a distributed knowledge system: A constructionist approach. *Strategic Management Journal* 17, no. S2 (1996): 11–25. doi:10.1002/smj.4250171104.
8. Stacey, Ralph D. The science of complexity: An alternative perspective for strategic change processes. *Strategic Management Journal* 16, no. 6 (1995): 477–95. doi:10.1002/smj.4250160606.

9. Bechky, Beth A. Sharing meaning across occupational communities: The transformation of understanding on a production floor. *Organization Science* 14, no. 3 (2003): 312–30.
10. Ben-Menahem, Shiko M., Georg von Krogh, Zeynep Erden, and Andreas Schneider. Coordinating knowledge creation in multidisciplinary teams: Evidence from early-stage drug discovery. *Academy of Management Journal* 59, no. 4 (2015): 1308–38. doi:10.5465/amj.2013.1214.
11. Heath, Chip, and Nancy Staudenmayer. Coordination neglect: How lay theories of organizing complicate coordination in organizations. *Research in Organizational Behavior* 22 (2000): 153–91. doi:10.1016/S0191-3085(00)22005-4.
12. https://en.wikipedia.org/wiki/Building_information_modeling.
13. https://cpd.uw.edu/transition-occupancy-t20.
14. Haas, Aurore. Crowding at the frontier: Knowledge brokers, gatekeepers, boundary spanners and marginal-intersecting individuals, May 2014. https://basepub.dauphine.fr//handle/123456789/13553.
15. Vlaar, Paul W.L., Frans A.J. Van den Bosch, and Henk W. Volberda. Coping with problems of understanding in interorganizational relationships: Using formalization as a means to make sense. *Organization Studies* 27, no. 11 (2006): 1617–38. doi:10.1177/0170840606068338.
16. https://en.wikipedia.org/wiki/Issue_tracking_system.
17. https://askamathematician.com/2011/10/q-what-is-the-three-body-problem/. Accessed July 12, 2018.
18. Coase, R. H. The nature of the firm. *Economica* 4, no. 16 (1937): 386–405. doi:10.1111/j.1468-0335.1937.tb00002.x.
19. Oliver Williamson | Department of Economics. Accessed January 23, 2019. https://econ.berkeley.edu/profile/oliver-williamson.
20. The Sveriges Riksbank Prize in Economic Sciences in Memory of Alfred Nobel 2009. NobelPrize.org. Accessed January 23, 2019. https://nobelprize.org/prizes/economic-sciences/2009/press-release/.
21. Elinor Ostrom. Econlib. Accessed January 23, 2019. https://econlib.org/library/Enc/bios/Ostrom.html.
22. Security Liability in an 'Assume Breach' World. Dark Reading. Accessed January 24, 2019. https://darkreading.com/partner-perspectives/f5/security-liability-in-an-assume-breach-world/a/d-id/1331100.
23. F5 Labs. Accessed January 24, 2019. https://f5.com/labs.html.
24. Boddy, Sara, Justin Shattuck, Debbie Walkowski, David Warburton. The Hunt for IoT: Multi-Purpose Attack Thingbots Threaten Internet Stability and Human Life. F5 Labs, October 24, 2018. https://f5.com/labs/articles/threat-intelligence/the-hunt-for-iot--multi-purpose-attack-thingbots-threaten-intern.html.
25. Ransomware. In *Wikipedia*, January 23, 2019. https://en.wikipedia.org/w/index.php?title=Ransomware&oldid=879846895.
26. What Is a Botnet? And Why They Aren't Going Away Anytime Soon | CSO Online. Accessed January 24, 2019. https://csoonline.com/article/3240364/hacking/what-is-a-botnet-and-why-they-arent-going-away-anytime-soon.html.
27. What Is a Distributed Denial-of-Service (DDoS) Attack? Cloudflare. Accessed January 24, 2019. https://cloudflare.com/learning/ddos/what-is-a-ddos-attack/.
28. Loryka. Accessed January 24, 2019. http://loryka.com/.

Additional Reading

Barley, Stephen R. 1986. Technology as an occasion for structuring: Evidence from observations of CT scanners and the social order of radiology departments. *Administrative Science Quarterly* 31: 78–108.

Bechky, Beth A. 2003. Sharing meaning across occupational communities: The transformation of understanding on a production floor. *Organization Science* 14, no. 3: 312–30.

Ben-Menahem, Shiko M., Georg von Krogh, Zeynep Erden, and Andreas Schneider. 2016. Coordinating knowledge creation in multidisciplinary teams: Evidence from early-stage drug discovery. *Academy of Management Journal* 59 (4): 1308–38. doi:10.5465/amj.2013.1214.

Bijker, Wiebe E. 1995. *Of Bicycles, Bakelites, and Bulbs: Toward a Theory of Sociotechnical Change. Inside Technology.* Cambridge, MA: MIT Press.

Bruns, H. C. 2013. Working alone together: Coordination in collaboration across domains of expertise. *Academy of Management Journal* 56 (1): 62–83. doi:10.5465/amj.2010.0756.

Carlile, Paul R. 2002. A pragmatic view of knowledge and boundaries: Boundary objects in new product development. *Organization Science* 13 (4): 442–55. doi:10.1287/orsc.13.4.442.2953.

Carlile, Paul R. 2004. Transferring, translating, and transforming: An integrative framework for managing knowledge across boundaries. *Organization Science* 15 (5): 555–68. doi:10.1287/orsc.1040.0094.

Carrillo, P., and P. Chinowsky. 2006. Exploiting knowledge management: The engineering and construction perspective. *Journal of Management in Engineering* 22: 2–10. doi:10.1061/(ASCE)0742-597X(2006)22:1(2).

Dossick, Carrie S., Gina Neff, Laura Osburn, Christopher Monson, and Heather Burpee. 2016. Technical boundary spanners and translation: A study of energy modeling for high performance hospitals. In *Working Paper Proceedings.* Cle Elum, Washington.

Foss, Nicolai J., Kenneth Husted, and Snejina Michailova. 2010. Governing knowledge sharing in organizations: Levels of analysis, governance mechanisms, and research directions. *Journal of Management Studies* 47 (3): 455–82. doi:10.1111/j.1467-6486.2009.00870.x.

Haas, Aurore. 2014. Crowding at the frontier: Knowledge brokers, gatekeepers, boundary spanners and marginal-intersecting individuals. In Rennes, France. https://basepub.dauphine.fr//handle/123456789/13553.

Heath, Chip, and Nancy Staudenmayer. 2000. Coordination neglect: How lay theories of organizing complicate coordination in organizations. *Research in Organizational Behavior* 22: 153–91. doi:10.1016/S0191-3085(00)22005-4.

Kaplan, Sarah, Jonathan Milde, and Ruth Schwartz Cowan. 2017. Symbiont practices in boundary spanning: Bridging the cognitive and political divides in interdisciplinary research. *Academy of Management Journal* 60 (4): 1387–1414. doi:10.5465/amj.2015.0809.

Law, John. 1992. Notes on the theory of the actor-network: Ordering, strategy, and heterogeneity. *Systems Practice* 5 (4): 379–93. doi:10.1007/BF01059830.

Leonardi, Paul M. 2012. *Car Crashes Without Cars Simulation Technology and Organizational Change in Automotive Engineering.* Cambridge: MIT Press.

Organizational Innovation and Change. internal-pdf://2008 Whyte_Digital Innovation_ final-2467412480/2008 Whyte_Digital Innovation_final.doc.

Orlikowski, W. J. 2007. Sociomaterial practices: Exploring technology at work. *Organization Studies* 28 (9): 1435–48. doi:10.1177/0170840607081138.

Pickering, Andrew. 1993. The mangle of practice: Agency and emergence in the sociology of science. *American Journal of Sociology* 99 (3): 559–89. doi:10.1086/230316.

Star, Susan Leigh, and James R. Griesemer. 1989. Institutional ecology, 'Translations' and boundary objects: Amateurs and professionals in Berkeley's museum of vertebrate zoology, 1907–39. *Social Studies of Science* 19: 387–420.

Whyte, Jennifer. 2008. Model Risk, Digital Innovation and Complex Organizations. In *Digital Challenges in Innovation Research Workshop*. Philadelphia, PA: National Science Foundation.

Chapter 6

IoT Systems Manageability

Why Do We Care About IoT System Manageability?

We care about IoT system manageability because in our resource-constrained institutional and city environments which incorporate a rapidly growing number of resource – consuming IoT systems, we want to select and implement IoT systems that have high manageability. These systems of high manageability have reduced impact and reduced demand on our limited institutional and city resources.

> We want the manageability of IoT systems to be critical criteria for IoT system selection and implementation.

We care about IoT systems manageability because in addition to the known costs of selection, procurement, deployment, and operation of IoT systems:

- Managing and operating IoT systems have nonobvious and nuanced costs across the institution or city
- Because there are an increasing number of IoT systems, these costs aggregate
- Increased interdependency between IoT and other systems increases systems complexity and costs
- Costs aggregate in an environment of constrained funding for staffing and support, particularly around institutional and city infrastructure
- Manageability of IoT systems has a direct impact on demand for limited city and institutional resources – that is, good IoT systems manageability reduces demand on limited resources

■ IoT systems that have higher manageability can decrease the transaction cost losses that occur between the many institutional or city organizations and departments involved with using, supporting, and/or maintaining the Systems
■ This ability to positively impact the delta between institution/city costs and available resources in the same domain also aggregates and can have a critical impact on the ability of a city or institution to manage its systems, deliver positive ROI, and not degrade its cyber risk exposure.

For example, consider an IoT system that mounts sensors (cameras, other) that detect and communicate traffic conditions and provide control input for traffic management across a city or institution. Some questions that impacts systems manageability include:

■ How difficult are the sensors of the system to install and replace?
■ Is there a known approximate rate of failure?
■ How many sensors can be out before system is considered substantially degraded – e.g., 20%? 10%, 5%, and 0% sensor failure okay?
■ How often will they need to be serviced?
■ What skill sets are required to service the sensor?
■ Is special access required?

Systems with unreliable sensors are in frequent need of service in challenging locations, such as the top of a city streetlight, impact system success and directly impact the cost of servicing and operating the system. This perception of failure is, of course, problematic. Also, problematic, though, is that this additional cost burden to operations and support pulls precious resources – time, staffing, and funding – from an already very constrained pool of resources. This, in turn, affects the long-term operability of the IoT system in question, *as well as* other IoT systems using the same pool of resources. This is an example of weak IoT systems manageability.

Taking the converse of the above example, a system with reliable sensors rarely in need of service reduces demand on the constrained resource pool. This effect, in turn, positively impacts the IoT system in question *as well as other systems reliant on the same pool of resources for support.*

Another example of an IoT system implementation aspect to analyze for manageability might be usability, configurability, and supportability of the server application software that manages, configures, and gathers data from the endpoints (IoT devices). Using the same traffic sensing/management system example above

■ Is configuring data collection and aggregation on the IoT system's control and data aggregation server intuitive to use?
  – Is there good documentation?
  – Is vendor support readily available?
  – Is there a peer group of other city and institutional users?

■ Is the data collection and aggregation application installed on a standard operating system?
  – Does the application require particularly unique or unusual operating system configurations?
■ How difficult is it to update or patch the application software?
  – How often does this occur?
  – Is system downtime required to patch/update software?

A problematic IoT system with low systems manageability might be one where the system is difficult and/or nonintuitive to configure and operate. Such software systems lend themselves to system dysfunction as well as increased likelihood of errors made in software/system configuration and operation. Similarly, an awkward installation process and/or requirements for unusual operating system configurations are time-consuming, can increase system downtime, and erode the support team's confidence in the system. Finally, in this example, difficult software updating or patching processes can lend itself to unnecessarily lost time. It can also manifest itself in non-updated and unpatched software systems if the updating/patching is too difficult and/or time-consuming to do.

The flipside, of course, is an IoT system with good manageability that has easily upgradeable and patchable systems, standard operating systems, and fairly standard operating system configurations.

Identifying and Analyzing Problematic IoT Systems Manageability Areas

Because these sociotechnical IoT systems are so complex, identifying risk areas and components affecting systems manageability can be daunting. Known problematic areas around systems manageability identified by intuition, or "gut feel," and/or experience are a critical component. However, because there are so many places for costly systems manageability problems and because there is not a lot of precedent for identifying these aspects in increasingly complex and novel IoT systems, intuition alone can miss some of these risk areas.

Another approach is try to be exhaustive with all of the possibilities of risk areas and systems manageability issues. This can be good because it has a higher likelihood of identifying all of the candidate risk/systems manageability areas in IoT systems for subsequent analysis. However, the downside is that, because it does identify so many areas, actual analysis in all of those areas can be impractical from a time and staffing point of view for most cities and institutions. Further, because aspects of these analysis processes need to be done repeatedly, perhaps annually or biannually for each system, an exceptionally long list of possibilities to analyze is even more problematic from an available resource point of view (Figure 6.1).

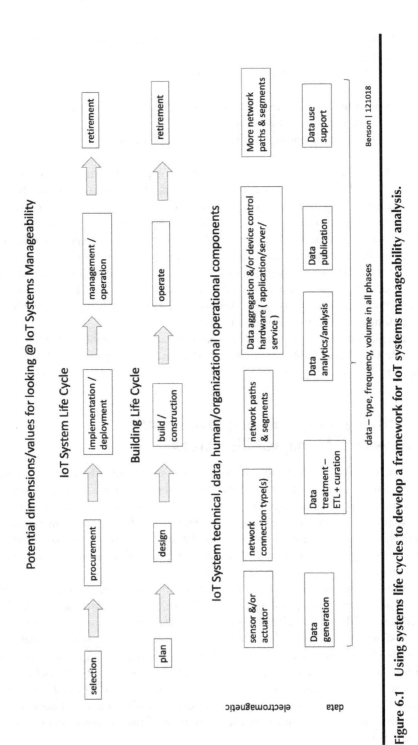

Figure 6.1 Using systems life cycles to develop a framework for IoT systems manageability analysis.

Because of these factors, I advocate a hybrid approach where (1) intuition/ "gut feel" and experience are combined with (2) a near-exhaustive list of possibilities. (Note: it is probably impossible to create an exhaustive list of possibilities because of the changing nature of components and interdependencies of these systems.) In this approach, the analysis starts with documenting the intuitive and experiential areas of systems manageability problems. Once documented, a deep list of possibilities and areas of problematic systems manageability is generated and reviewed at a high level for additional areas that may trigger further attention and analysis.

One way to generate this large list of possible problematic areas is to construct a framework around

- The life cycle of buildings/facilities
- The life cycle of IoT systems
- The components of IoT systems – device, network, supporting hardware, data, and others.

The Life Cycle of an IoT System

IoT systems have five broad phases within their life cycle:

- IoT system selection
- IoT system procurement
- IoT system implementation/deployment
- IoT system operation, management, and support
- IoT system retirement.

During the IoT system selection phase, ostensibly an institutional or city need has been identified. From there, a process is executed, or set of processes, where that need is articulated for the many constituents within the city or institution whether they be city managers and citizens within a city or students, faculty, staff, and public within a large Higher Education institution such as a university or other. Further, ideally, all city or institution organizations or departments involved in the support, maintenance, and operation of the system have the opportunity to weigh in on the selection. The process may include sending out requests for information (RFIs),[1] requests for proposals (RFPs),[2] scheduling vendor meetings (early versions of which often referred to as "dog and pony shows"), analysis of RFPs for alignment with requirements, other reviews and processes, and a final decision on the desired system.

Unfortunately, at this point in the evolution of selecting and implementing IoT systems, this structured, thorough, and definitive approach just described may be more the exception than the rule. More commonly, the system selection decision may be driven more by shiny sales pitches made to only a fraction of the stakeholders affected and promises to solve unarticulated problems.

During the IoT system procurement phase, in the course of system acquisition, additional processes will be executed with the proposed IoT system vendor. Contracts will be drafted, reviewed, and negotiated. Funding sources will be identified and, as needed, packaged and aggregated for acquisition purposes. Phased timing of payments contingent on particular benchmarks or events may be established. Ideally, the IoT system will also be analyzed for compliance with any existing city or institution purchase requirements. That said, the novelty, complexity, and rapidly evolving nature of many IoT systems can make this analysis difficult to align with requirements established for older, more "traditional" technology systems.

During the implementation or deployment phase, the IoT system is installed in the city or institution. This is a critical phase and most institutions and cities do not yet perform it well. It involves identifying requirements, system component interdependencies, resource alignment – e.g., who does what – resource scheduling, and others. This phase can also have a substantial project management component. Ideally, much of this identification and alignment will have been identified earlier during the selection and/or procurement phases, though currently that is often not the case (Figure 6.2).

During the operations, management, and support phase, the IoT system is actually executed and operated to deliver the intent of its purchase. Almost all stakeholders are impacted during this phase. During this phase, assumptions get tested, oversights expose themselves, and performance – good or bad – is delivered. Assumptions around expectations of what data delivers, actual required support

Figure 6.2 Complex physical and complex IT systems in modern buildings. (Photo courtesy McKinstry.)

resources, costs of support, vendor availability, operator proficiency, and many others get aired out. During this phase, the efficacy and thoroughness of the deliverables requirements also reveal themselves.

ON OPERATIONAL TECHNOLOGY (OT)

A critical consideration of both the implementation/deployment phase and the operations/management phase of the IoT system is skill sets available for the implementation and subsequent operation. IoT systems devices deployed in built environments require both traditional IT skills *and* traditional trades and engineering skills. In the words of a leading IoT services provider for Building Automation Systems and energy management systems,

> "... we are often talking about physical assets so *resolving an issue often takes both data analysis and a wrench*, few people have both skill sets. Identifying faults but not fixing them just undermines overall program objectives ...."
> *Hendrik Van Hemert, Pacific Northwest Regional Director, McKinstry*[3]

The Operational Technology (OT) skill sets blend two very different cultures with different histories, traditions, and perceptions. For example, the two cultures have very different perceptions and experiences with time. The traditional IT aspect views time in terms of months, days, hours, and minutes as IT professionals manage IT systems in data centers and office spaces. Construction and facilities management professionals, on the other hand, can view time in terms of decades as they are experienced in constructing and then operating buildings that last for decades (Figure 6.3).

Similarly, IT professionals are used to changing systems configurations with great frequency as software updates with new features and functionality regularly come out as well as frequent software patches to address a rapidly changing and growing array of cyber security vulnerabilities. Facilities management personnel often manage large utility distributions to buildings such as heat, electricity, water, and others. Disruptions to these systems are very noticeable and can be critical. For this reason, facilities management professionals are naturally motivated not to change these systems unless absolutely necessary.

Neither of the above approaches are wrong. They evolved from very different backgrounds to address different needs. The critical thing to note is that they are now coming together at a very rapid pace. Those professionals that have this blended skill set of OT – e.g.,, and that have the skill and qualifications to work with high voltage in physically hazardous environments *and* configure computer-based and networked meters, sensors, and actuators are in short supply and high demand.

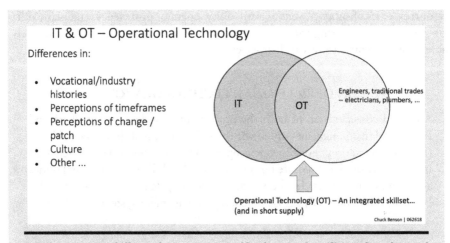

Figure 6.3 OT skill sets have aspects of both IT and traditional trades and engineering.

The impact of the short supply and high demand of these OT skill sets is that large and numerous IoT systems deployments in institutions can suffer when these skill sets are not available. The opportunity of poor or improper configuration is high – and that directly impacts the institution's ROI and cyber risk posture.

Finally, during the retirement phase of the IoT systems life cycle, the IoT system is deemed to no longer have utility and the system can be shut down and retired. Another reason for retirement of the IoT system is that the building, facility, or space in which it is deployed is being demolished or perhaps completely retrofitted. Because IoT systems are still new, there are even fewer examples of IoT systems in the retirement phases. What occurs or does not occur in the retirement phase can be very impactful and potentially problematic however.

When a deployed IoT system is either explicitly deemed not of value (i.e., an organizational decision is made) or implicitly (i.e., people simply quit using the system and/or the data generated by it), *the deployed system still exists*. The fact of an explicit or implicit condition of non-usability *did not turn off, power down, decommission, or uninstall the system*. Ideally, all of the system components are thoughtfully and thoroughly decommissioned. For example, any server hardware or virtual machines are powered down and destroyed (e.g., disks wiped), any supporting SaaS or other cloud servers/services have their provider's deletion/destruction processes executed – including for associated data, and client workstations are wiped of applications and associated data.

Perhaps most importantly, the decommissioning and uninstallation of tens, hundreds, thousands, or more ideally needs to occur so that there are not that many forgotten networked computers still operating – and still adding to the enlarged attack surface – on the institution's or city's network (or that of its partners and neighbors). This decommissioning/uninstallation work of these numbers of geographically destroyed sensors or actuators is, of course, very expensive. And the odds are high that most cities and institutions will not physically de-install the sensors and actuators, though some efforts may be made to deactivate them remotely.

The substantial components and activities associated with the different phases of the IoT systems life cycle can provide a critical analysis tool for IoT systems for cities and institutions.

The Life Cycle of a Building

Buildings and facilities in which many IoT systems are deployed and operate also have a life cycle. A simplified version of a building life cycle includes these phases:

■ Building planning
■ Building design
■ Building construction
■ Building operation
■ Building retirement.

During the planning (and preplanning) phases of a building's life cycle, project objectives and deliverables are identified, and feasibility studies are conducted. A project plan and supporting plans are created and matured. Standards requirements and performance requirements are also identified and documented. Budgets, risk analyses, and control plans are also developed.

During the design phase, architectural aspects of the building, the engineering aspects of the building, and the technical aspects of the building are considered and choices for implementation are made. Incorporating work from the planning phase, the Design Intent is also established and communicated.

During the building's construction phase, materials are delivered, skilled tradespeople and teams are engaged and coordinated, and the building is constructed. Also, during this phase, discoveries made by construction contractors that conflict or are otherwise not in alignment with the building's Design Intent are communicated by the contractor to the architect/designer (Figure 6.4).

During the operations phase of a building's life cycle, the building is occupied by users/tenants, and supported, maintained, and operated by facilities personnel. Operating and maintaining a building includes repairs and alterations by skilled trades such as plumbers, electricians, carpenters, and other skills. Important to note, many of these roles are becoming increasingly complicated as the mechanical, electrical, and plumbing (sometimes called MEP[4]) systems become increasingly complex. Many of these systems are, in part or in whole, IoT systems.

Figure 6.4 Teams of skilled professionals engage in building construction. (Photo courtesy McKinstry.)

A Framework for Analysis

The phases of the respective IoT life cycle and building life cycle can make up two axes of an analysis framework. A third axis for the analysis framework can be the components of an IoT system. IoT systems have a spectrum of devices, networks, supporting systems, data considerations, and others. For example, some IoT systems components and aspects include:

- IoT endpoint device or actuator
- Supporting networks
  - Wired
  - Wireless
  - Different network segments.
- Data components existing in various states across time and location
- Supporting hardware such as
  - Data aggregating servers
  - Device controlling/configuring hardware.

This framework can be visualized as a three-dimensional coordinate system with lists of each of the above forming the axes (Figure 6.5).

With such a construction, every permutation of IoT life cycle phase, building, or facility can be delineated and considered. At each of the permutations, an analysis of systems manageability could occur which would include:

- Impacts on staffing
- Project/process disruptive impacts
- Deleterious fiscal impacts.

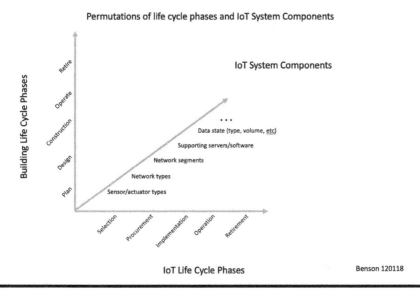

Figure 6.5 Three-dimensional framework consisting of IoT system life cycle, building life cycle, and IoT system components.

For example, scanning through the three-axis diagram can suggest areas of consideration such as

- What are *network considerations* during *IoT system selection* that occurs during a *building's planning* phase?
- Similarly, for a *building's design phase* – What are *network considerations* during *IoT system selection* that occurs during the building's design phase?
- What are *considerations for the sensors* that must be installed and deployed during a building's *construction phase?*
- What are potential concerns/issues of an *IoT system's operation* during a *building's operating phase* for *data generated?*

In practice, actually analyzing every possibility may well be impractical for most cities and institutions given all of the potential permutations. That said, a deep list such as this can be reviewed at a high level to see if particular permutations – in combination with intuition and experience – trigger additional analysis for issues for areas of IoT systems manageability (Figure 6.6).

To exhaustively run through all of the permutations of IoT life cycle phases, building life cycle phases, and IoT system components, we can run a simple Python script on an online Python interpreter such as https://tutorialspoint.com/execute_python_online.php. Of course, if you're comfortable with basic Python, you can run it yourself on any desktop and laptop (or even some IoT devices themselves such as a Raspberry Pi!).[5]

```
# generate permutations between IoT Systems life cycle phases, building life
cycle phases,
# and IoT Systems components -- to assist with analyzing problematic IoT
Systems manageability

iotlifecycle =        ['selection','procurement',
                      'implementation/deployment',
                      'management/operation','retirement']

buildinglifecycle =   ['planning','design','construction',
                      'operate','retirement']

iotsystemscomponents = ['sensor/actuator', 'network connection type',
                      'network path', 'aggregation/control device',
                      'more network path(?)', 'data tx/ETL/curation',
                      'data analysis/analytics', 'data publication',
                      'data use support']
count = 0
for i in iotlifecycle:
        for b in buildinglifecycle:
                for c in iotsystemscomponents:
                        print count
                        print '''If I am in the IoT System %s phase & '''%i
                        print '''the building is in the %s phase, '''%b
                        print '''what are the considerations of
                        manageability/support of the %s? \n'''%c
                        count+=1
```

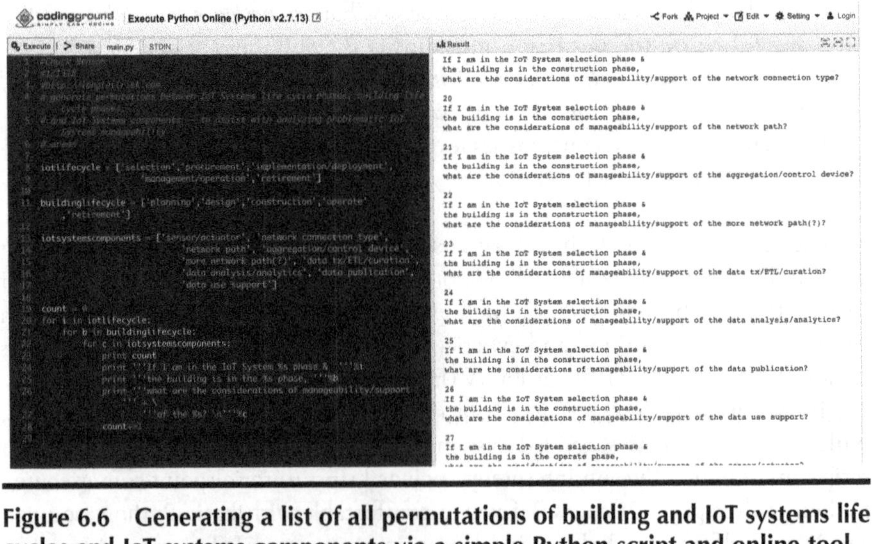

Figure 6.6 Generating a list of all permutations of building and IoT systems life cycles and IoT systems components via a simple Python script and online tool – https://tutorialspoint.com/execute_python_online.php.

This will yield many permutations of life cycle phases and IoT system components. The number of permutations is equal to

```
# of building phases X # of IoT System phases X # of IoT
System components
```

For the example that I use, that number is 224 permutations. Of course, not all of these permutations will be useful for analysis. However, by scanning through the output at a high level, it can be used to prompt ideas on problematic manageability areas that pure intuition and/or experience do not (Figure 6.7).

Samples from the results include:

```
2
If I am in the IoT System selection phase &
the building is in the planning phase,
what are the considerations of manageability/support of the
network path?
```

One example might be a research institution that needs to meet a regulatory or compliance need for research in a sensitive area in a new or refurbished building or other space, such as a laboratory space. The need may call for an environmental monitoring system – an IoT system – deploying hundreds of sensors across a geographically distributed area and across multiple networks, subnets, or other network segments. Questions to ask about networks in this IoT system selection phase and building planning phase could include:

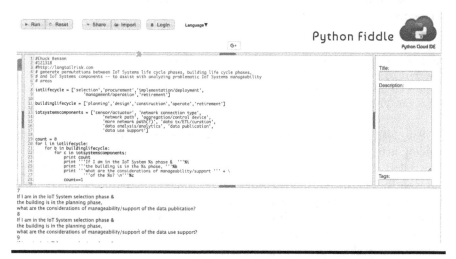

Figure 6.7   Executing the same code with another online Python interpreter.

- Exactly what networks does the institution anticipate installing these sensors and supporting hardware on?
  - If the networks or network segments do not exist yet, what upstream networks will they connect to?
- Does the central IT organization know that this installation requirement is coming?
- How will the vendor, the IoT system provider, update the software and/or firmware on those sensors?
  - Are there firewall rules to address?
  - Are there access controls to navigate?
  - Who is managing either of the above?
    - Will that staffing be available?
    - Is that staffing in an institutional organization that is "on board" with the effort? That is, is the organization providing the staffing supportive? Do they know?
- Do the sensors and/or any client applications and/or supporting servers have to "phone home" in order to perform? If so, are additional firewall rules, aka firewall "holes," needed?
  - Do additional permissive firewall rules impact other existing (or future) systems on the network?

These are just some of the questions that different organizations might be interested in asking and participating in the building planning phase and IoT system selection phase. Discussing these and other issues during the building's planning phase can save substantial grief at the time of implementation and subsequent operation. (The number at the top is the number of that particular permutation of building life cycle phase, IoT system life cycle phase, and IoT system component).

```
16
If I am in the IoT System selection phase &
the building is in the design phase,
what are the considerations of manageability/support of the
data publication?
```

Similarly, using the same example, during the IoT system selection phase while the building is in its respective design phase, there are good questions to ask around data collection, analysis, and publication needs. For example, it is increasingly common for new or refurbished building owners to share building energy usage data with compliance organizations, tenants, and possibly the public. Some of this sharing is via reports and web dashboards, while some via real-time display on "video walls" of building performance. If this is difficult to do, it directly impacts systems manageability and supporting resources.

78
If I am in the IoT System procurement phase &
the building is in the operate phase,
what are the considerations of manageability/support of the
data analysis/analytics?

With coordination and mutual training, the procurement office can play a substantial role in the IoT systems manageability assessment by, as a condition of procurement, questioning that the deliverables/objectives of this system are well understood by the purchaser of the system and the IoT system supplier.

153
If I am in the IoT System management/operation phase &
the building is in the construction phase,
what are the considerations of manageability/support of the
sensor/actuator?

This permutation of an IoT system in its operation phase while the building is in construction phase may not have a lot of utility for analysis currently. However, that could change in the future as IoT systems, and particularly "future old" IoT systems have increased prevalence in cities and institutions.

162
If I am in the IoT System management/operation phase &
the building is in the operate phase,
what are the considerations of manageability/support of the
sensor/actuator?

If the IoT system is operating and the building is occupied and operating, what are the considerations that affect IoT systems manageability? Using the same example as above of an environmental monitoring system for a research institution with sensitive components, questions on IoT systems manageability (and the related loss and costs for poor systems manageability) can be system performance:

- Is the IoT system providing expected results for all constituents?
  - Operational needs
  - Reporting.
- Are the costs of support and maintenance within expectations?
  - Do we *know* what the actual costs incurred are?
    - For example, do we know all of the institutional or city constituents – e.g., maintenance organizations, central IT, local IT, tenants – that are bearing the costs of the IoT systems operation?

Other permutations of IoT system life cycle, building life cycle, and IoT system components include:

166
If I am in the IoT System management/operation phase &
the building is in the operate phase,
what are the considerations of manageability/support of the
more network path(?)?

167
If I am in the IoT System management/operation phase &
the building is in the operate phase,
what are the considerations of manageability/support of the
data tx/ETL/curation?

168
If I am in the IoT System management/operation phase &
the building is in the operate phase,
what are the considerations of manageability/support of the
data analysis/analytics?

211
If I am in the IoT System retirement phase &
the building is in the operate phase,
what are the considerations of manageability/support of the
network path(?)?

An Initial IoT Systems Manageability Assessment Process

In general terms, a partial IoT systems manageability assessment process/analysis criteria could include at least the following:

- Sensor
    - Ease of configuration
    - Ease of installation
    - Ease of operation
    - Ease of servicing
    - Ease of replacement
    - Power requirements?
    - Data quality
    - Self-reporting health status?
- Aggregating/controlling device/server
    - Ease of Configuration
    - Ease of Operation
    - Ease of Replacement
    - Data quality

- Built in connections to external data sources, e.g.,
  - Microsoft SQL Server, Oracle database server, MySQL etc.
  - Cloud service, e.g. AWS, Azure, Google
  - Or are connections non-standard/proprietary, requiring special configuration or coding to integrate?
- Does it provide any early treatment or curation of data?
  - For example, does it handle missing values? Or just forward them as is?
■ Network
- How much work is involved in deploying sensors on this particular set of networks, network segments, firewalls, etc.
- Sensor network protocols
  - Common, standard? Or unusual?
■ Data processing
- Does data arrive in standard formats?
- Is data made available in standard formats? Or is translation/interpretation required?
■ Data publication
- Easy?
- Actionable?
■ Analysis
- Proprietary?
- Readily interfaces with other analysis tools such as Tableau,[6] Microsoft Power BI,[7] Azure Analytics,[8] AWS IoT Analytics,[9] or others.
■ How difficult is it to configure and operate
- The data aggregation server and software application
- Client applications
- Sensors and actuators.
■ Does the data feed from the IoT system readily integrate with other data systems, IoT, or otherwise?
■ Precedence
- Has the IoT system in question been deployed before?
- Where? Can you talk to the client about their selection, implementation, and operation experience with this particular IoT system?
- How similar/dissimilar is it to your network?
- Do they have a similar set of constituents?

Recap

To revisit the original premise, IoT systems can provide substantial social and business benefit, but only if thoughtfully selected, implemented, and operated. These systems are themselves complex and exist in complex sociotechnical environments of constrained resources. The combination of the system loss that accompanies these complex sociotechnical systems and that the systems are operating in resource-constrained environments is problematic and directly impacts ROI as well as the city or institution's cyber risk exposure. Resource constraints include but are not limited to:

- Available staffing
- Skill sets required
- Scheduling challenges
- Disruptions
- Trust.

Manageability of an IoT system can immediately and directly positively impact that systems loss. More manageable systems can reduce the staffing required to operate, manage, and support the IoT system. This also has the effect of being less disruptive and reducing resource scheduling overhead and challenges. IoT systems that are more manageable also have the effect of enhancing, or at least not degrading, trust between the institution/city and the IoT systems provider/vendor.

Suggested Reading

1. Request for Information. In *Wikipedia*, November 27, 2018. https://en.wikipedia.org/w/index.php?title=Request_for_information&oldid=870893787.
2. Request for Proposal. In *Wikipedia*, November 20, 2018. https://en.wikipedia.org/w/index.php?title=Request_for_proposal&oldid=869799029.
3. Van Hemert, Hendrik. Interview with Hendrik Van Hemert, Pacific Northwest Regional Director - Technical Services, McKinstry, December 17, 2018.
4. Mechanical, Electrical, and Plumbing. In *Wikipedia*, November 28, 2018. https://en.wikipedia.org/w/index.php?title=Mechanical,_electrical,_and_plumbing&oldid=871087259.
5. Raspberry Pi Foundation. Raspberry Pi — Teach, Learn, and Make with Raspberry Pi. Raspberry Pi. Accessed January 24, 2019. https://raspberrypi.org.
6. Tableau: Business Intelligence and Analytics Software. Tableau Software. Accessed January 24, 2019. https://tableau.com/.
7. Power BI | Interactive Data Visualization BI Tools. Accessed January 24, 2019. https://powerbi.microsoft.com/en-us/.
8. Azure Analytics Services | Microsoft Azure. Accessed January 24, 2019. https://azure.microsoft.com/en-us/product-categories/analytics/.
9. AWS IoT Analytics Overview - Amazon Web Services. Accessed January 24, 2019. https://aws.amazon.com/iot-analytics/.

Chapter 7

IoT Systems Vendor Relations and Vendor Management

The relationship between the city or institution, that is the consumer or user of an IoT system, and the provider of that system is critically important and has a direct impact of the manageability of the Internet of Things (IoT) system. As we saw in Chapter 6, this systems manageability in turn has a direct impact on the success of the IoT systems implementation – measured in terms of both Return on Investment (ROI) and cyber security/cyber risk posture.

As discussed earlier, IoT systems stakeholders (whether they know that they are stakeholders or not) can span broadly across organizations within the city or institution with often unclear ownership. This fact coupled with general lack of deep precedence for selection, implementation, and operation of IoT systems, the rapid evolution and ongoing novelty of these systems, and the intermingling and increasingly fuzzy network boundaries between the institutional consumer and IoT systems vendor/provider means that the relationship between them plays a particularly heavy role.

Because this relationship is so important to IoT systems manageability and hence a strong determinant of success of the IoT system implementation, choosing the vendor/provider is very important (and vice versa), as well as managing the relationship after selection of system and provider is made (Figure 7.1).

Ideally, we would like this relationship to exist with maximum added value to the institution or city and simultaneously with minimal systems loss – which

Causal relationships on IoT System ROI & Institutional/City Cyber Risk

Benson | 120118

Figure 7.1 Client–vendor relationships impact manageability of IoT system, substantially impacting likelihood of implementation success.

reflects in ROI and cyber security/cyber risk posture across the institution or city. To that end, we would like that interface to be as efficient and productive as possible.

Asymmetry of Roles in IoT Ecosystem

While the city or institution pursuing the deployment of an IoT system and the potential IoT system vendor/provider are both agents in the IoT ecosystem,

> the market nature of bringing this institution user of an IoT system – the institution or city – together with the vendor/provider of the IoT system, creates an asymmetry.

In the market, the city or institution "gets to shop" for different IoT systems as well as different IoT systems providers. In turn, the different alternatives of IoT systems and the respective vendors/providers of them work to avail themselves to the buyer and compete with other providers in the process.[1] Because of this, while IoT system vendor/provider considerations are also covered, the number of considerations and recommendations in the following section is skewed higher for the city or institutional customer identifying an IoT system for purchase and deployment.

Know Thyself

The most important thing for a city or institution to do, prior to selecting, acquiring, and deploying an IoT system, is to understand itself and know what it wants, what its objectives are. Shakespeare's "to thine own self be true,"[1]

---

[1] Mature IoT vendors and providers are also selective regarding the client institutions and cities that they take on, but the bulk of the market pressure is that institutions and cities shop for IoT systems providers versus the other way around.

Plato's expansion of the notion to "an unexamined life is not worth living,"[2] and Sun Tzu in *the Art of War* tells us to "know yourself"[3] as a necessary ingredient of winning one hundred battles. All of these reflections, observations, and directives reflect the importance of knowing what we are and knowing what we are made of – to include the systems that we have that can largely define us.

These hold true also when an institution or city, which are already complex sociotechnical systems, seeks to implement an IoT system for its constituents in the interest of adding social or business value (or both) for its constiutents. In the course of implementing and operating an energy management system, air quality monitoring system, traffic safety system, or potentially value adding IoT system, the institution also takes on risk by further adding complexity. Before engaging with potential providers of the IoT system(s) needed to accomplish these objectives,

> the institution or city needs to understand who its varied constituents are, what the objectives are of deploying these additional sociotechnical systems in its spaces, and what resources it has to implement, manage, and operate the systems (as well as the ones that it already has).

Early in the process, while it makes sense for city or institution to engage with various IoT systems vendors/providers to explore the possibilities of various new IoT systems technologies and services, it is important to not move too deeply in the selection, purchase, deployment, and management phases of the system, until the "know thyself" question is answered.

Some of the areas that the institution or city needs to ask itself include:

- Who are the constituents that will be affected by the purchase and deployment of this new sociotechnical system?
  - To what degree might they be impacted?
    - What is the expectation that the perception of the system investment and deployment is positive?
    - What is the expectation that the perception of the system investment and deployment is negative?
  - What data is produced by the system?
    - How will the data be used? How will it be analyzed?
    - How will the data be shared?
    - Across the varied constituencies, how might the data be interpreted?
    - How could the data be used maliciously or otherwise other than intended?
- What problem is the city or institution trying to solve with the selection, purchase/investment, and operation of this new system?
  - What is the desired value added? Economic benefit? Risk reduction? and others
  - Has that been formalized, documented, and communicated?

- How is the ROI calculated?
  - What is criteria for measuring return? Was value created?
    - Revenue increases?
    - Cost savings elsewhere?
    - Risk reduction?
  - How will the total cost of ownership costs be assessed?
    - How will costs incurred across multiple institutional or city organizations or departments be collected?
    - How will this cost data be aggregated?
    - Who will do this work? What people, departments, etc.?
- How will the institution or city measure changes in cyber security/cyber risk for the institution or city?
  - What does the institution's current cyber risk profile look like?
  - Who will do this work? Chief Information Security Officer (CISO)? Others?
    - Is staffing available or planned for to do this work?
- How long will the System operate? 5, 10, 20 years?
  - What will be required to decommission the system when desired?
    - What to do with the deployed endpoints/devices/sensors?
    - Who will do the work? Do future budgets plan for this decommissioning work?
- Are there existing institutional/city policies and procedures that guide these endeavors of IoT system acquisition and implementation?
- Are there contract templates and/or addenda for IoT systems implementations and operations?
- Are there any historical examples of IoT systems implementations and operations that the city or institution has undertaken?

For cities, constituents can be the many different neighborhoods, sectors, and locale within the city's limits. Further, constituents can also include the many different occupations and types of business within the city. Other demographics can include age, income, race/nationality, family status, and others. The city will need to look at as many of these as it can in the context of this new system that could impact all constituents. For institutions, such as universities, constituents can include all of the above as well as additional roles or types such as students, faculty, staff, public visitors, and others.

It is important that the city or institution knows what problem that it is trying to solve and exactly how the IoT systems under consideration could possibly solve those problems. Often times, cities and institutions do not articulate this problem to be solved well, much less document, and communicate among its constituents. This has a number of ramifications:

1. It degrades the institution's ability to analyze options of both the IoT systems themselves and the vendors that provide them

2. It prevents analysis of contract performance during the System's operations phase.

Without some reasonable clarity on that value looks like, or return looks like, for the IoT system, it is impossible to calculate ROI which is one of the two components of IoT systems implementation success.

Regarding ROI, it is also important to know how actual costs of acquiring, implementing, and operating the system will be calculated. Reminding ourselves that

The R in ROI = actual value received less actual costs incurred

is critically important and a core criterion of the success of an IoT system implementation.

Calculating these costs is typically not straightforward and can be difficult to accomplish. Because support and operation of IoT systems span many different institutional or city organizations and departments, costs associated with these support activities can also be spread out across organizations and departments. Often departments and organizations within the institution or city do not identify and calculate costs in the same manner. This is because of history and precedent, but also because departments have very different things to account for. Each individual department can have substantially different approaches and business processes for cost accounting. This can make identification of all costs and aggregation of the same particularly difficult.

For example, costs of a large environmental monitoring and control system for a laboratory in a research university may include central IT for network services with staffing costs, operations costs, amortized networking hardware costs, and others. Support for the system likely also involves staffing and management from the facilities management organization for endpoint/device/sensor support such as an array of air quality sensors embedded in the ceiling. It may also involve data center costs for hosting data aggregating servers that support data collection for subsequent analysis. Actual users of the system, in this case researchers, may incur unexpected costs in the form of staff time used for generating and collating regulatory reports and responses.

The CISO's office as well as the risk management office may be involved to assess recently introduced cybersecurity exposure from the system. These costs can be felt by the institution in terms of staff time and/or opportunity cost (i.e., other things not getting done).

Also, early in the "getting to know one's self" process, identifying any policies, procedures, relevant compliance or regulatory documents, and/or institutional guidance is also important prior to entering into the engagement with the IoT system vendor.

For a successful IoT systems implementation, it is important that the city or institutional purchaser/acquirer of IoT systems understands its capabilities and capacities for supporting and operating the system as well as understanding the

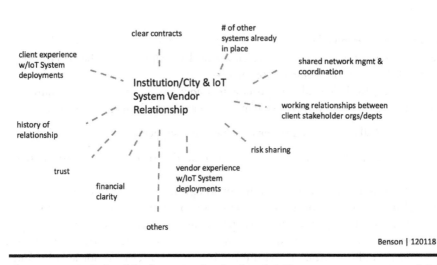

Figure 7.2 **Relationship between institutional IoT system consumer and IoT system provider is critical – and there are many aspects to consider and manage.**

impacts of the implementation to the city's or institution's cyber risk posture and cyber security needs. To effect this,

> the city or institution *must raise the bar regarding its expectations of the IoT systems provider.* Without this expectation-setting and articulation of specific IoT system deployment and operational needs, there is little pressure for IoT systems vendors and providers to change and strengthen their deliverables.

Some of these changes are expensive for the provider, and there are natural motivations to not incur the expense. However, if cities and institutions broadly increase these expectations of IoT systems deliverables, IoT systems providers and vendors will need to adapt in order to be competitive (Figure 7.2).

Courtship of IoT Systems Customer with IoT Systems Vendor/Provider

The relationship between the institutional/city customer and the IoT systems vendor can be a bit like dating, particularly during the early phases. There is a lot of excitement and a lot of potential (to go either way), no one really knows anybody, opportunities for miscommunication and misperceptions are many, and the actual long-term outcome is unknown.

To mitigate some of the challenges of this early courtship phase, it is helpful for both parties to ask some questions to help structure the relationship and provide early tests for the relationship. I'll start with questions from the institution/city IoT systems customer to the IoT systems vendor/provider and then go the other direction and suggest some prospective questions that the IoT systems vendor may want to ask the client.

Early dating questions for the IoT systems customer to the potential IoT systems provider/vendor:

- How long has your IoT system been on the market?
- How many other IoT systems have you deployed?
  - Who are those customers?
  - Can we speak to them about the experience?
  - How similar were those deployment environments to ours?
  - Are they all active?
- Is your IoT system modular? (How flexible are you?)
  - Or does the entire end-to-end process and system need to be purchased and supported?
  - Sensors/actuators
    - Can your sensors/actuators work with other systems?
- Data aggregation
  - What the available are data formats? How can I get system data?
    - Standard? Proprietary? Home grown? Prepackaged? Customizable?
  - Does data aggregation occur on premise with hardware or VM?
    - Or cloud service?
    - What are costs of supporting hardware and software services?
- Data processing, treatment, curation, etc.
  - Is your approach to this phase proprietary? Open? Uses other data products/services?
- Data analysis
  - Proprietary?
  - Uses other industry standard tools, e.g., Tableau?
  - Flexible? Customizable?
- Data publication
  - How do you make data available for end user?
    - Data interfaces for customer-defined approaches?
    - Pre-built dashboards? Predefined Key Performance Indicator (KPI)'s?
      - If so, are these particular pre-built dashboards and KPI's useful for my institution's or city's defined objectives?
- Cyber security
  - Do you have published internal policies and procedures for good cyber "hygiene" practices? Can you share them with me?

■ Risk
  – Do you have templates or existing contract language regarding risk sharing for this implementation?
    • Both parties benefit from successful implementation.
  – Identify who is liable for what if something goes wrong.

Red flags for the IoT systems customer during the early dating phase:

■ Vendor has limited IoT system deployments of similar type and scale that the customer is considering
■ Vendor won't share (or doesn't have) contact information for other clients
■ Vendor doesn't ask questions about nature of customer's network
  – For example, segmentation, access control, and firewall policies
  – Related, vendor assumes customer's network is like theirs or very simple and homogenous.
■ Vendor insists that their product has a "secret sauce"[4] that makes it better than competitors but doesn't explain why
■ Vendor organization has no facilities management experience in leadership or staff.

Early dating questions for the IoT systems vendor/provider to the potential IoT systems customer:

■ What problem are you trying to solve for your institution or city?
  – Have you reviewed this with all of your stakeholders?
  – Anybody (stakeholder) not on board with the idea of purchasing and deploying this new IoT system?
  – Can we meet the key stakeholders?
■ What are your expectations of this product and service?
■ Have you selected, deployed, and managed other IoT systems in the past?
■ Do all of your stakeholders that will be providing support for the IoT system know that they will be providing support?
■ How many IoT systems are you already operating?
  – How many endpoints/devices/sensors/actuators do you currently support?

Red flags for the IoT systems vendor/provider during the early dating phase:

■ Potential IoT systems customer has little to no experience selecting, procuring, implementing, and operating IoT systems for their city or institution
■ Representation from key support organizations not present in early interactions, e.g.,
  – Central IT/networks not represented, facilities management organization not represented, CISO's office unaware of IoT system, and others.

- Appearance of excessive current or historical friction between organizations/ departments in the city or institution
- Institution or city does not have clearly defined objectives and expected outcomes from the IoT system under consideration
- Potential customer's existing systems do not seem to follow reasonable standards or best practices.

Relationship Management Phase – The Marriage and "knotworking"

Once the IoT systems institutional or city customer enter into a contract with the IoT systems provider, the relationship is not unlike a *marriage*:

- There is apparent alignment of purposes and objectives
- Promises are made
- Bumps in the alignment of purposes and objectives are discovered or developed over time
- Resolution and negotiating (and hard work) activities occur
- Some issues are resolved and some not so much.

While analysis in the dating phase is critical for setting up the best footing, much of the work between provider and customer comes in this relationship management phase.

Researcher Yrjo Engestrom introduces the term "knotworking" in his paper, *Activity theory as a framework for analyzing and redesigning work,* based on research in the healthcare setting and others. He indicates the importance of the concept as "New forms of work organization increasingly require negotiated "knotworking" across boundaries." This notion is also very apropos in this complex of IoT systems selection, delivery, and operation in this complex of institutions and cities, networks, and IoT systems vendors/providers.

Engestrom refers to this as an "emerging type of work" and continues

> The notion of knot refers to the … distributed and partially improvised orchestration of collaborative performance between otherwise loosely connected actors and activity systems. A movement of tying, untying and retying together seemingly separate threads of activity characterizes knotworking. The tying and dissolution of a knot of collaborative work *is not reducible to any specific individual or fixed organizational entity as the centre of control. The centre does not hold. The locus of initiative changes from moment to moment* within a knotworking sequence.[5]

He extends on the "co-configuration" work of Victor and Boynton[6] where "[the firm] builds an ongoing relationship between each customer-product pair and the company … brings the value of an intelligent and "adapting" product … the customization becomes continuous [and] never results in a "finished" product. Instead, a living, growing network develops between customer, product, and company."[6]

Engestrom further describes knotworking as an ongoing iterative process of co-configuration. He summarizes the criteria of co-configuration[5] as

1. Adaptive product or service
2. Continuous relationship between customer, product/service, and company
3. Ongoing configuration or customization
4. Active customer involvement
5. Multiple collaborating producers
6. Mutual learning from interactions between the parties involved.

There are a number of phrases, terms, and descriptors in this discussion of knotworking that ring true for the IoT ecosystem of vendors, cities and institutions, and sociotechnical systems:

Engestrom's Knotworking or Co-configuration Descriptor	Author's IoT Ecosystem Parallels and Examples
"across boundaries"	IoT systems span many organizational boundaries, both internally and with/between IoT systems providers/vendors
"distributed"	Sensors, aggregating servers on premise or in cloud, data processing in client workstations, cloud, or on-premise servers, and others – substantial distribution of system components
"partially improvised orchestration of collaborative performance between otherwise loosely connected actors and activity systems"	The spanning of IoT systems across multiple organizations involving multiple providers for different system component parts often require supporting organizations to learn and adapt "on the fly"
"tying, untying and retying together seemingly separate threads of activity"	The relationships and iterative coordination, execution of work, and adapting between multiple institution/city orgs and departments and one or many providers of IoT systems or subcomponents over the many parts of the IoT system feels very much like tying, untying, and retying

(Continued)

Knotworking or Co-configuration Descriptor	IoT Ecosystem Parallels and Examples
"[the] knot of collaborative work is not reducible to any specific individual or fixed organizational entity as the centre of control. The centre does not hold. The locus of initiative changes from moment to moment"	The multiple parties, roles, activities, interdependencies, and the iterative and adaptive nature of operating and sustaining IoT systems in the complex organizational structures of institutions and cities definitely has mobile, shifting sense of control. There is no one hierarchy or command and control structure that exists or suffices
Continuous customization that "never results in a 'finished' product. Instead, a living, growing network develops between customer, product, and company"	Very true of IoT systems – the supporting entities, the iterative continual adaptation is a theme in supporting IoT systems across multiple institutional/city orgs, multiple vendors, and multiple IoT system components
All of the co-configuration components – adaptive; continuous relationship between customer, product/service, and company; ongoing configuration/customization; active customer involvement; many collaborative producers; mutual learning across parties	Essential components of IoT systems selection, deployment, and management/operation success for city and institutional systems consumers

The city or institution and IoT system vendor/provider relationship is something like a marriage. There is a courtship phase, sometimes it works, and sometimes it does not. Following that, there is a long-term relationship phase, sometimes it works, and sometimes it does not.

It is critical for the institutional or city IoT systems consumer to recognize and acknowledge that potential IoT systems vendors/providers that do not want to engage in, or have a history of engaging in, *a knotworking process* of a long-term relationship will likely have a product or service that is doomed to fail within the city or institution's context.

Ability, capacity, and willingness – none of which are trivial to either party – to work the "knotworking" aspect are critical to successful IoT systems implementations for cities and institutions.

Suggested Reading

1. To Thine Own Self Be True - Meaning, Origin, and Usage. Literary Devices, May 29, 2015. https://literarydevices.net/to-thine-own-self-be-true/.
2. The Unexamined Life Is Not Worth Living. In *Wikipedia*, January 23, 2019. https://en.wikipedia.org/w/index.php?title=The_unexamined_life_is_not_worth_living&oldid=879876137.
3. The Internet Classics Archive | The Art of War by Sun Tzu. Accessed January 27, 2019. http://classics.mit.edu/Tzu/artwar.html.
4. Definition of SECRET SAUCE. Accessed January 27, 2019. https://merriam-webster.com/dictionary/secret+sauce.
5. Engestrom, Yrjo. Activity theory as a framework for analyzing and redesigning work. *Ergonomics* 43, no. 7 (2000): 960–74. doi:10.1080/001401300409143.
6. Victor, Bart, and Andrew C. Boynton. *Invented Here: Maximizing Your Organization's Internal Growth and Profitability.* Boston, MA: Harvard Business School Press, 1998.

Chapter 8

Templates for Institutional and City IoT Systems Planning and Operations

---

With all of the many components, agents, departments and organizations, systems of systems, social technical systems, and other aspects of Internet of Things (IoT) systems deployed in institutions and cities, figuring out where to start and how to proceed with implementing, supporting, and living with IoT systems can be a challenge. A helpful and structured approach for doing this is to use the phases of an IoT system's life cycle to organize what tasks and questions should be asked when and by what groups, organizations, and departments.

For example, the initiator of the acquisition and implementation of the IoT system can be one of several different people, departments, or organizations. It could be a building tenant, or Principal Investigator (PI) in a university, a facilities management organization, or others. Ostensibly, the initiator of the IoT system acquisition, regardless of which department or organization it is, has a particular problem that it is trying to solve or value that it is trying to add through the purchase, deployment, and operation of the IoT system.

To review, there are five phases in an IoT system life cycle:

- Selection
- Procurement
- Implementation/deployment
- Operation/management (the value generating phase)
- Retirement.

While cities and institutions can vary among themselves regarding how their functions are distributed across organizations and/or departments, the following functions/organizations are fairly common across large institutions and cities:

- Planning and budgeting
- Finance
- Central IT (usually includes networking services)
- Facilities management
- Capital development
- Risk management
- Distributed IT (IT support organic to some departments/organizations)
- Chief Information Security Officer
- Building/facility tenants, academic researchers/practitioners, and others
- Privacy official/office
- Procurement office
- Senior administrative leadership (university presidents, chancellors, etc.), city mayors.

Other possible organizations to consider within the institution or city includes:

- Law enforcement/police department
- Business continuity/disaster recovery
- Donors/benefactors
- University regents, trustees
- Internal audit.

For the template to follow, I'll use a subset of the above. For your institution or city, adjust, add, subtract, or otherwise modify the list to best suit your model.

Ask, Task, Communicate

Within each of the IoT life cycle phases, one or more organizations will have tasks to accomplish, information to share, and questions to ask. Importantly, some of this information sharing and these questions to ask are from one organizations/department in the institution or city to another within the institution or city. This is where that coordination and cooperation begins to occur and that will also need to be nurtured as the process continues (Figure 8.1).

For each phase, we'll identify some questions, tasks, and information to share and which groups should be asking, tasking, communicating during each phase. The list will not be exhaustive, and there are some that may not apply for every institution or city. It does, however, provide a basis or starting point for analysis and

IoT Systems Planning & Operations Template for Institutions & Cities

Questions, tasks, communication:

IoT System Life Cycle Phases:

	selection	procurement	implementation	operation	retirement
questions to ask by/to which dept or org	• Lorem ipsum dolor sit amet, usu no facilisi	• Clita alterum offendit duo in	• Ad moderatius interpretaris vix Vulputate intellegam	• Per et nibh recusab honestatis. Fugit primis te sed, partem	• ius eu alia idque dicant, has ne natum assum
Tasks to accomplish by/to which dept or org	• ius eu alia idque dicant, has ne natum assum	• Per et nibh recusab honestatis. Fugit primis te sed, partem	• Ad moderatius interpretaris vix Vulputate intellegam	• Clita alterum offendit duo in	• Lorem ipsum dolor sit amet, usu no facilisi
information to share between depts/orgs	• ius eu alia idque dicant, has ne natum assum	• Lorem ipsum dolor sit amet, usu no facilisi	• Clita alterum offendit duo in	• Ad moderatius interpretaris vix Vulputate intellegam	• Per et nibh recusab honestatis. Fugit primis te sed, partem

Benson | 121518

Figure 8.1   What to ask, task, communicate between organizations within the IoT system life cycle.

planning for almost all institutions and cities. So, for each phase in the IoT systems life cycle, there will be Ask-Task-Communicate actions that will be performed by some subset of all of the groups/organizations within the institution or city.

IoT Systems Selection Phase

This phase is one of the larger phases and will benefit from at least some involvement and input from virtually all organizations or departments within the institution or city. This list is not exhaustive. Add and modify as best fits your institution and city. Examples include the following:

Ask section:

■ All groups/organizations ask the initiator of the IoT system acquisition the purpose and intent of the system
■ Effort coordinator/champion or designated project manager (PM) for the IoT system implementation (explained in "Task section") – ask the city or institution these questions:
  – How many IoT systems are already deployed?
  – How many devices/endpoints are already deployed?
  – What is the success of the deployed systems?
    • Is it known? Has it been calculated?
      ■ Return on Investment (ROI)?
      ■ Degradation of cybersecurity/cyber risk profile?
    • How well are the existing systems being managed?
■ The effort coordinator/champion should ask (possibly via the effort PM) central IT group:
  – What information it needs and what concerns and questions it has over prospective system
■ Central IT group should ask
  – What, if any network segmentation, will be required?
  – What is geographic distribution of deployment?
    • Covers many subnets? Routing instances?
    • Do new routing instances need to be created to support this IoT system?
    • What kind of access control is needed?
  – How much IP address space?
  – What are anticipated network bandwidth needs?
  – Does the IoT systems vendor provide any systems support software and/ or tools? For example,

- Device enumeration?
- Device health?
- Device diagnostics?
- Others.
  - Is new hardware/software required?
    - Are there new wired Ethernet "drops" needed?
      - Where?
      - Costs to deploy?
    - Are new Wireless Access Points needed?
      - Where?
      - Costs to deploy?
- Effort coordinator/champion ask existing IoT systems stakeholders what support experience has been like for supporting existing systems?
  - Has vendor been able to support?
  - Have internal resources such as central IT, facilities management, and others been able to support existing IoT systems? (If not, questions should be raised about implementing another IoT system in need of management).
- Are there privacy concerns regarding the potential new IoT systems implementation?
- Facilities management group should ask:
  - What are anticipated support efforts?
    - By trade or shop? Electricians, plumbers, engineers?
  - Where has this system been deployed before?
    - Who are the existing customers?
      - What have their facilities support experiences been?
  - How will support time be accounted for the effort?
- The effort coordinator/champion and/or the finance organization should ask:
  - How will support costs be accounted for across multiple supporting organizations?
    - Will they use similar accounting/costing approaches?
    - How will these different costs be aggregated or "rolled up" across all of the system cost-bearing units/departments?
  - These questions are critical for determining the cost portion of the ROI equation.
- Effort coordinator/champion asks <u>Who pays for what?</u>
  - Who pays for original IoT system purchase?
    - Who pays for support and maintenance costs?
    - Who pays for supporting services, e.g., supporting servers and cloud services?

- Estimate effort levels, e.g., staffing hours per month, and associated costs for each supporting organization
- Ensure supporting organizations participate in this process.

Task section:

- Identify an effort coordinator or champion. Ideally, this is a senior institutional or city leader. This senior leader, in turn, can designate a PM that ensures questions, information sharing, and tasks/work are accomplished at each phase. The PM or designee can also collect, organize, and synthesize the information of all phases.
- Involving as many groups/organizations as possible – *review, iterate, and articulate the objective of the system.*
  - Though often a challenging task, without clear broadly understood system objectives, it is impossible to manage subsequent performance expectations.
- Determine key participants for supporting the system. For example,
  - Will facilities management organization be involved?
    - If so, which parts of facilities management?
  - Will central IT be involved?
    - If so, which parts?
  - Will distributed IT be involved?
    - If so, which parts?
  - Will building tenants, occupants, end users be involved?
- Develop and distribute an RFI (request for information) if possible
  - Articulate requirements of desired IoT system
  - Ensure as many groups/orgs as possible have opportunity to provide input into requirements
    - Different organizations, groups, constituencies can have different expectations of the system.
      - Can be particularly true regarding expectations of data generated by the IoT system.
  - Include boilerplate template for IoT systems RFI standards.
- Review RFI responses with team
- Develop and distribute RFP (request for proposal)
  - Develop RFP articulate needs and objectives identified across institution & city
  - Include boiler plate template for IoT systems RFP standards.

Communicate section:

- As much as possible, inform affected, impacted groups – whether support organizations or citizens in the case of a city or students/faculty/staff in the case of a higher education institution of new potential IoT system.

IoT Systems Procurement Phase

Ask section:

■ Procurement office will ask what are the funding sources for this IoT system purchase?
  – What budgets and budget numbers?
  – Has approval been obtained?
■ Procurement organization should ask facilities management and/or capital development group if there are existing institutional/city design guides that require adherence
■ Are there regulatory or compliance guidance or mandates that the purchase must be aligned with?
■ Are there institutional or city government policies within which the IoT systems must align?
■ Are there risk-sharing contracts or contract addenda to identify or attach to the primary contract with the IoT systems vendor/provider?

Task section:

■ Begin contract development/review with IoT systems provider
■ In contract, include "who will do what" section regarding system implementation and operation
  – Will IoT system vendor or institutional/city internal resources configure and install the endpoint IoT devices?
  – Or both?
■ In contract, include hard description of deliverables of endpoints (IoT devices themselves), supporting server hardware (whether local, on-premise, cloud service(s), or some combination of all of the above)
■ In contract, if vendor is performing all or part of the IoT device/endpoint implementation/deployment as part of required deliverable, include the following at a minimum:
  – Vendor will deliver documentation that includes:
    • Physical location of all deployed devices
    • IP address of all deployed devices
    • MAC addresses of all deployed devices
    • Firmware version for all deployed devices
    • Hardware version for all deployed devices
    • Software version for all deployed devices
    • Model number of all deployed devices
    • Patch plan (firmware/software update) plan for all deployed devices.

- In contract, specify and enumerate what ports/services are required to be open for device operation.
  - Further, specify that all nonessential IoT device ports/services are disabled
  - Document this in post-implementation documentation (aka "as-built"[1]documentation).
- In contract, include that vendor will deliver post-implementation commissioning and testing
  - Include results of nmap port scans[2] of all devices and date/time of scan
  - Include results of traceroute execution for each endpoint (from a known institutional or city network location).
- In contract, include specific language for ongoing patch plan for IoT devices/endpoints
  - Include mechanism for patching
  - Frequency of patching
  - Who is responsible for patching, e.g., IoT system provider or customer.
  - Include any additional costs related to patching
- In contract, specify risk-sharing agreement between institutional/city consumer and provider
  - Financial risk, e.g., underperforming ROI for the IoT system
  - Cyber security risk for the IoT system.
    - Who is responsible for cyber incident stemming from IoT system applications, devices, configurations, and operations?
- Iteratively, as necessary, seek clarifying guidance from stakeholders via the IoT system acquisition coordinator/champion
- As necessary, work with finance organization to identify and encumber funds for purchase and subsequent support of the system

Communicate section:

- Keep IoT system acquisition coordinator/champion informed of the contract development and negotiation.

IoT Systems Implementation Phase

Ask section:

- As early as possible, facilities management and/or capital development organizations request from central IT organizations a range or sets of ranges of IP addresses on the institution or city network
- As early as possible, reaffirm whose responsibility it will be to support and operate different aspects of the system. For example,

- Who will support the endpoints/devices? Facilities management? Vendor? Other?
- Who will support and operate any on-premise supporting hardware, software, and/or services such as data aggregation and central device control.

Task section:

- Facilities management organization and central IT organization collect "as-built" documentation from the vendor
- Facilities management organization develop early draft of any preventative maintenance (PM) work needed
- Central IT organization cross-check actual IP address implementation with the range of IP addresses issued/provided earlier in the process. Document any provided, but unused addresses

Communicate:

- As early as possible, facilities management and/or coordinator/champion share intended deployment locations of all IoT devices with central IT
  - This allows central IT to plan network address usage.
- As early as possible, facilities management and/or coordinator/champion reaffirm network segmentation requirements with central IT
  - This allows central IT to plan technical and staffing resources.
- Communicate to all stakeholders any changes to deployment plan
- Communicate to all stakeholders any surprises/unexpected things learned during deployment.

IoT Systems Operations and Management Phase

Ask section:

- IoT systems coordinator/champion, iteratively ask stakeholders – tenants, PIs, facilities management, etc. – their impression of performance of the deployed system.
  - What is working/working well?
  - What is not working/working well?

Task section:

- IoT system coordinator/champion schedule regular review meetings for all stakeholders to gather and review, discuss, problem solve, and document issues with operation and support of the system. Earlier in the operation, the meetings should be more frequent. If/as system performance improves and/

or becomes more predictable, the frequency of the review meetings can be adjusted. Also, it is possible that the number of participants in the review/troubleshoot meetings can become smaller.

- Anticipate at least biweekly review/troubleshoot meetings for the first several months after deployment
- For larger and more complex IoT systems, this will require regular, ongoing review/discussion (perhaps years and perhaps permanently)
- Deployed IoT systems that will continue to scale, e.g., continually adding more sensed points, will likely require ongoing, regular, fairly frequent review/troubleshoot/planning meetings.

■ Central IT and facilities management organizations meet periodically to review network configurations, performance, network change planning, system growth planning, and others. Perform regular audits of subset of network, even if informal.

Communicate section:

■ All groups/organizations communicate to IoT system coordinator/champion impressions of performance.

- This contributes to system champion's ability to discern ROI for the system (or delegate of the champion).

■ All groups/organizations communicate to IoT system coordinator/champion (or delegate), cost information incurred in the course of support of the systems

- For example, the facilities management organization should track and communicate monthly staffing hours spent in support of the system.

■ Communicate to all stakeholders any changes to original operational plan.
■ Communicate to all stakeholders any surprises/unexpected things learned during operation and management of system.

IoT Systems Retirement Phase

Ask section:

■ IoT system coordinator/champion query all stakeholders on ongoing utility of IoT system.

Task section:

■ Decommission system. Physically uninstall and/or remotely disable all sensors and actuators

- Run network tests to confirm no response
- If not physically removed, perform visual inspection if possible that power has been removed from each device (to include onboard battery if present and removal feasible).

Communicate section:

■ Communicate to all previous stakeholders that system is being decommissioned.

Considerations for Phases of Building Life Cycle

The previous discussion holds true for virtually all IoT systems deployed in built environments at any time. There can be some special cases and considerations, though, that vary with the phase of the *building* life cycle:

Building planning phase:

■ How many IoT systems will be procured, implemented, and operated in the new building?
■ Who will be the IoT system coordinators/champions for each of these systems?
■ Are there interdependencies between these multiple IoT systems or other systems?
■ What sort of vendor coordination will be required between multiple systems?
■ Who, within the institution or city, will oversee, monitor, and mediate communication between multiple IoT systems providers?
■ Are there desires and/or opportunities to integrate and leverage data between multiple systems?
  – If so, what internal or external resources are available to support this data integration, analysis, visualization, output, etc.?

Building design phase:

■ What points of intersection will multiple IoT systems have within the building or facility?
  – Physical? Network? Data?
■ If there are desires and/or opportunities to integrate and leverage data between multiple systems, what technical resources should be identified and tasked to develop those data opportunities?

Building construction phase:

■ Are there physical conflicts with device deployment from multiple IoT systems?
■ Do sensors and actuators from different systems conflict with each other? For example, does a room fan or blower conflict with the operation of an air quality sensor?

- Are there conflicts with cable runs to multiple sensors from multiple different systems?
- Are there conflicts with identifying devices and supporting network and power cabling from multiple different systems?
- Are there Wi-Fi conflicts between sensors and actuators?
- Is there sufficient Wi-Fi bandwidth for all sensors or actuators using Wi-Fi for their networking?

Building operation phase:

- Will implementation of the IoT system after the building is already operating interfere with building operation?
- Will implementation of the IoT system increase or disrupt any existing IoT system interdependencies?
- How many additional endpoints/devices will the new IoT systems add to the building or facility?

Building retirement phase:

- If a building is to be retired, do IoT system endpoints/devices need to be removed?
- Do embedded IoT devices in the building contain residual privacy or otherwise sensitive information?
- If the building is to be rendered unoccupied, but not demolished, are there assurances that all devices have been removed from accessible networks and power removed?

Boilerplate for IoT System RFIs and RFPs

In developing RFIs and RFPs for IoT systems, there are a number of common aspects to be considered and that can be applied regardless of type of IoT system. The following can be used as the basis for an addendum to an RFI or RFP or can be embedded directly into the RFI or RFP language for the IoT system.

Example RFI and RFP language for IoT systems acquisitions:

IoT System Vendor X, Please Indicate your Ability to Perform the Following:	Can Perform? Yes/No	Can Provide Historical Examples? Yes/No
Provide post-installation "as-built" documentation of • System architecture drawing of completed IoT system deployment		

(*Continued*)

IoT System Vendor X, Please Indicate your Ability to Perform the Following:	Can Perform? Yes/No	Can Provide Historical Examples? Yes/No
• Location of all IoT devices/sensors/ actuators • Firmware version of all IoT devices/ sensors/actuators • Hardware version of all IoT devices/ sensors/actuators • Software version (as applicable) of all IoT devices/sensors/actuators		
Develop, document, execute patch plan for all IoT devices/sensors/actuators		
Develop, demonstrate, and execute plan or schema for ensuring all default login user and password information is changed and recorded in a secure manner		
Develop, demonstrate, and execute plan or schema for ensuring all unnecessary ports/services are disabled from all IoT devices/sensors/actuators		
Develop, demonstrate, and execute plan or schema for a comprehensive commissioning plan for entire IoT system to include endpoints/devices, supporting on-premise hardware and software, cloud-based services, hybrids, and expected results from data		

Conclusion

The IoT life cycle Ask-Task-Communicate activities as well as the template and boilerplate documents presented here are not exhaustive. They do, however, capture some of the essential elements and considerations for IoT systems implementations and operations in the built environment. The reader is encouraged to use these as starting points and to add, subtract, and modify as needed to most effectively apply to the environment and situation at hand.

By applying a structured approach such as that presented here, or a derivative thereof, we can increase the IoT systems manageability at each phase of the IoT system life cycle. This is particularly true for the longest phase of IoT systems operation and management.

By enhancing the manageability of these systems, we can reduce the loss – in terms of time, funding, resources, and other – that always occurs when implementing complex sociotechnical systems in the built environment.

The mitigation of this loss is of particular importance in the resource-constrained environments of institutions and cities. Systems loss mitigation work directly impacts IoT systems manageability and, in turn, the likelihood of IoT systems implementation success.

Suggested Reading

1. Riddell, Tyler. *What Are 'As Built' Drawings & How to Make Them*. eSUB Construction Software, June 14, 2017. https://esub.com/built-drawings-make/.
2. Nmap: The Network Mapper - Free Security Scanner. Accessed January 27, 2019. https://nmap.org/.

Chapter 9

Strategy Implementation

Internet of Things (IoT) systems are complex sociotechnical systems for which institutions and cities have little precedent for selection, procurement, implementation, and management. Resources and skill sets for their implementation and management are typically in short supply, and organizational and system complexities create a challenging situation for which we are poorly equipped as institutions and cities. As discussed, these challenges include:

- The complexity of the systems
- The organization-spanning nature of the systems
- The insidious systems loss and drain on limited resources caused in part by multiple transactions and interactions across many organizations
- The dependencies on existing technological infrastructure (such as networks)
- The interdependencies between IoT systems
- The rapidly growing number of IoT systems (and the concomitant growth in the number of IoT devices)
- Resource-constrained environments
- Complex relationships between institutional IoT system consumer and the IoT system vendor/provider.

All of these contribute to a demanding environment on dwindling resources. Left unmanaged, they contribute to increased likelihood of failed Return on Investment (ROI) and increased likelihood of degraded cyber risk/cybersecurity posture for the institution or city – ROI and changes to cyber risk posture being our measure of success/failure of IoT systems implementations.

The *manageability* of an IoT systems has direct impact on its success or failure because *more manageable IoT systems draw less on the constrained staffing, funding, time, skill set resources that institutions and cities have available* to them.

Further, the manageability of a singular IoT system also impacts the likelihood of success or failure of other institutional IoT systems because IoT systems implementation and support resources tend to come from some of the same resource pools.

Because there are a number of factors that cities and institutions cannot control regarding IoT systems implementations, IoT systems manageability becomes one of the key factors in the success of singular IoT systems as well as the institution's or city's portfolio of IoT systems.

Portfolios of IoT systems within cities and institutions deployed in their resource-constrained environments are not unlike the tragedy of the commons problem introduced by William Forster Lloyd[1] in 1833, popularized by Garrett Hardin[2] in the late 1960's, and elucidated and expanded upon by Elinor Ostrom[3] for which she received the Nobel Prize in 2009.

The general idea of the tragedy of the commons concept is that a single person, entity, or actor drawing upon a common resource pool can negatively impact all of the other people or actors dependent upon the same pool by seeking to optimize use of the pool for themselves.

For our IoT systems portfolio example, it's not that a single actor – IoT system in this – case seeks to optimize itself, but rather that the system's inherent manageability is poor and therefore the system is excessively costly to manage, or attempt to manage. This excessive cost, stemming from poor system manageability, diminishes resources available for other systems in the city or institution's IoT systems portfolio.

Therefore, choosing an IoT system and IoT system vendor/provider using systems manageability as primary criteria for selection and implementation is essential for the institution or city. Because the city or institution has limited ability to change other factors of the problem such as lack of experience and precedent in systems implementation and management, constrained resources, system interdependencies, and others, the importance of the controllable or influenceable factor of selecting for more manageable IoT systems is reinforced.

Four-Part Risk Mitigation Strategy for IoT Systems Implementations in Institutions and Cities

A four-part strategy can be deployed within institutions and cities to mitigate risk – both ROI and cyber – stemming from IoT systems (Figure 9.1):

- Policy
- Education, awareness, and assistance
- Know yourself, know the threat
- Interorganizational coordination.

A 4 Pillar Approach to Institutional IoT Risk Mitigation Strategy

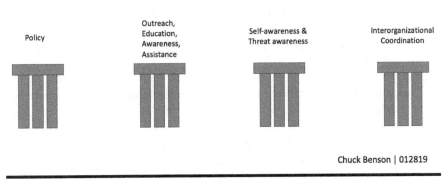

| Policy | Outreach, Education, Awareness, Assistance | Self-awareness & Threat awareness | Interorganizational Coordination |

Chuck Benson | 012819

Figure 9.1   One approach to institutional IoT systems risk mitigation.

Policy

Institutional or city policy and enforcement of the same are essential for an IoT systems risk mitigation strategy. While policy cannot address everything and enforcement has costs and challenges, it does provide a picture of institutional goals, objectives, and expectations.

Some considerations for institutional IoT policy components include:

- Default logins and passwords changed on all IoT devices
  - California's recent Senate Bill 327,[4] approved in September, 2018, is one of the first to institutionally mandate something similar. While some of the language is vague, such as "reasonable security," the bill does state, "preprogrammed password is unique to each device manufactured" and that "the device contains a security feature that requires a user to generate a new means of authentication before access is granted to the device for the first time."
- The IoT systems vendor/provider specifically designates and documents the exact ports and services that each IoT device is required to have open to achieve its intended function
  - Background: Ports and services on computers or IoT devices are small computer programs that are usually always running to provide some kind of designated connectivity service for the computer or device. The services have names, and the corresponding ports have numbers. For example, there may be a remote access port/service (such as Telnet/Port 23, SSH/Port 22, Windows Remote Desktop/Port 3389), a web interface (usually Port 80, Port 443, or Port 8080), and industry protocol like Modbus[5] (Port 502) or Bacnet[6] (Port 47808) to work with Building Automation Systems.
  - All ports/services provide opportunity for benign/appropriate use as well as malicious use.

- Some ports are essential for operation and they should be open.
- Ports that are not required should be disabled so that those venues for attack or misuse are not available.
- This is reducing the "porousness" of the attack service as discussed in Chapter 3.

■ Upon completion of system implementation, the IoT systems vendor or provider will deliver "as-built" documentation that documents at a minimum:
  - The location of all devices
  - The hardware model number of each device
  - The serial number of each device
  - The firmware version number of each device
  - Any associated software versions on each device
  - The Media Access Control (MAC) address of each device
  - The IP address of each device and its type (i.e., static or dynamic)
  - Software version of server software, Software-as-a-Service services, and client application software
  - Attestation that default login has been changed according to a defined and documented schema and that provided to appropriate institution or city representative
  - Attestation that all nonessential ports/services have been disabled. Test results of nmap[7] scans or similar to demonstrate the same
  - A system architecture diagram depicting relationships of IoT devices, server software, client application software, data flows, and interactions with other IoT systems in institution's or city's IoT systems portfolio.

■ A patch plan, aka firmware or software update plan, for all devices is in place and tested. IoT systems vendor/provider provides actual demonstration of patch plan mechanism

■ Departments, organizations, and divisions within a city or institution will notify any potential stakeholders, e.g., central IT, and facilities management organization of acquisition, deployment, and management of IoT system
  - Could have criteria associated with it, such as "any system deploying ten or more devices," or similar.

Education, Awareness, and Assistance for Institutional or City Stakeholders

The organizations, departments, and divisions within an institution or city need to know what IoT systems are, what inherent risks are, and what criteria are for IoT systems implementation success. Most of these would also be very appreciative of and grateful for guidance on how to proceed with IoT systems, particularly the selection, procurement, implementation, and operations/management phases. This education, awareness, and assistance can take various forms to include, but not limited to

- Meetings, presentations, discussions on what IoT systems are, and considerations of different aspects of the IoT systems life cycle
  - Benefits of meetings and discussions include:
    - Development of shared language of IoT systems and the contexts in which they may be acquired and deployed
    - Development of trust that spans organizations. This has long-term benefits for IoT systems acquisition, deployment, and operation.
- Printed and online descriptive collateral on what IoT systems are and considerations of different aspects of the IoT systems life cycle
- Checklists and guidance documents for IoT systems acquisition, deployment, and operations
  - Importantly, in addition to providing helpful steps and guidance regarding IoT systems and their life cycle phases, they also serve as boundary objects[8] that help with coordination and communication across organizations within a city or institution.

Know Yourself, Know the Threat

To rephrase, Sun-Tzu's, "know your enemy, know yourself" guidance, we'll say, "know yourself, know the threat." In the IoT systems in institutions and cities context, it is more important to know ourselves, particularly what we control, what we have limited control of, and what we don't control at all. We cannot hope to reasonably manage our IoT systems, enhance ROI, and mitigate systems risks if we do not know ourselves, what we have, and how successful we are at managing that.

While we will never have complete knowledge about our institution or city, we should always strive to know more (within the bounds of our resources). Example of IoT systems, supporting technology infrastructures, and organizational knowledge that we would like to have include:

- Our supporting network architecture
  - How big is it?
    - How fast is it growing?
  - What are the existing network segments?
  - Planned network segments?
  - Portion that is wired?
  - Portion that is wireless?
    - Traditional wireless

- IoT-specific wireless protocols such as Bluetooth Low Energy,[9] NB-IoT,[10] 5G, and many others.
  - What organizations, departments, and projects are associated with which networks and network segments
  - Who manages various network segments?
    - To what degree are they managed?
    - Who, if any, is accountable for devices on those segments?
- What IoT devices are already on the network?
  - Which segments?
  - Do we expect them?
  - Do we know who owns them?
  - Do they have understandable behavior on the network?
  - Can the rate of growth on network segments be estimated?
  - Which devices do we know something about?
    - Such as MAC addresses, intent, and behavior.
  - Approximately what percentage of devices on the network do we know very little or nothing about?
- Organizations
  - What organizations have current investments in and/or manage IoT systems?
    - Facilities management groups?
    - Building occupants?
    - Central IT groups?
    - Distributed IT groups?
    - Research groups (in university example)?
    - Teaching groups (in university example)
- Interdependencies
  - What interdependencies between IoT systems exist?
  - What new or planned IoT systems will create new interdependencies?
  - Network interdependencies?
  - Data interdependencies?
  - Systems management resources drawing from shared resource pools.

These are must some of the things about which we would like to learn more about our institutions and cities. To accomplish this, the institution or city will want to have some sort of IoT device/system discovery process to address this knowledge gap.

It is important to note though that while there is a direct mitigation component that responds to problematic IoT implementation discoveries on the city's or institution's network, the effort focus should be on the top four components of the strategy – policy, education/awareness, institutional self-knowledge, and organizational coordination. It can be tempting to try to mitigate every problematic device discovery, but because of the large numbers of deployed devices, spending the resources to mitigate every single one will distract from the strategic approach which ultimately will be more effective.

System Ownership and Interorganizational Coordination

As discussed in Chapter 2, IoT systems span many organizations across the institution or city requiring coordination, cooperation, and often collaboration to facilitate systems implementation and management and operations success. And as discussed in Chapter 5, this coordination, cooperation, and collaboration are often easier said than done. Because of this

> *IoT systems implementations need a champion to deploy and implement the system. Similarly, deployed IoT systems operations need an owner within the institution or city that can provide ongoing oversight to its operation.*

An IoT system being implemented needs a single party to coordinate the implementation across the many organizations and departments that will need to interact at various moments in time over the course of the implementation. This can be facilitated through a project manager that represents the IoT system implementation champion. Without such oversight, the implementation will be uncoordinated, and opportunities for unmet goals across various constituencies are high.

Deployed IoT systems that are in the operations phase need a system owner or system coordinator. There needs to be a single coordinating party, perhaps supported by a team that borrows time from other tasks, that provides IoT system management oversight. This oversight includes analyzing whether operational objectives of the system are being met, analysis of IoT systems vendor deliverables and contract performance are being met, and coordination of support costing for the system (possibly across multiple departments). Further, this oversight includes partnering with cybersecurity and cyber risk professionals and analyzing changes to the institution's or city's cyber risk profile as a result of implementing and operating the IoT system. Without such specific oversight, determining the implementation success of the IoT system, in terms of ROI and cyber risk, is virtually impossible.

IoT systems in our institutions and cities are here to stay. They offer the potential for substantial social and economic benefit. However, due to the complexity and vast numbers of computing devices of the systems, they also offer substantial potential for lost investment and greatly increased cybersecurity exposure. Identifying, planning for, and resourcing these systems across their respective life cycles can greatly mitigate that risk.

Suggested Reading

1. William Forster Lloyd. In *Wikipedia*, December 2, 2018. https://en.wikipedia.org/w/index.php?title=William_Forster_Lloyd&oldid=871667111.
2. The Garrett Hardin Society. Accessed January 28, 2019. https://garretthardinsociety.org/.
3. Elinor Ostrom. Econlib. Accessed January 23, 2019. https://econlib.org/library/Enc/bios/Ostrom.html.
4. Bill Text - SB-327 Information Privacy: Connected Devices. Accessed January 28, 2019. https://leginfo.legislature.ca.gov/faces/billNavClient.xhtml?bill_id=201720180SB327.
5. Modbus Tutorial from Control Solutions. Accessed January 22, 2019. https://csimn.com/CSI_pages/Modbus101.html.
6. BACnet - The New Standard Protocol. Accessed January 22, 2019. http://bacnet.org/Bibliography/EC-9–97/EC-9–97.html.
7. Nmap: The Network Mapper - Free Security Scanner. Accessed January 27, 2019. https://nmap.org/.
8. What Is Boundary Object | IGI Global. Accessed January 28, 2019. https://igi-global.com/dictionary/boundary-object/2777.
9. The Basics of Bluetooth Low Energy (BLE), EDN. Accessed January 13, 2019. https://edn.com/5G/4442859/The-basics-of-Bluetooth-Low-Energy–BLE–.
10. Narrowband – Internet of Things (NB-IoT). *Internet of Things* (blog). Accessed January 13, 2019. https://gsma.com/iot/narrow-band-internet-of-things-nb-iot/.

Appendix A: Some IoT Systems Vendor Management Considerations for Higher Education Institutions

Internet 2 CINO IoT Systems Risk Management Task Force April 2017

Purpose of Document

This document is intended to provide different organizations within Higher Education institutions with items to consider as they engage with Internet of Things (IoT) Systems vendors at the different phases of selection, procurement, deployment, and management. For example, these items/talking points can be used within the request for information, request for proposal, procurement, contract negotiation, deployment, and management stages. Different organizations within an institution will have different interests in the process, and some organizations will have intersecting/overlapping interests with other institutional organizations.

It is acknowledged that IoT systems are selected, acquired, and deployed by Higher Education institutions through multiple paths. Systems may arrive through Principal Investigators (PIs and their laboratories), through planning and budgeting departments, facilities management groups, capital development organizations, central IT, distributed IT groups, and multiple vendors and subcontractors.

The more historical acquisition approach of selection, acquisition, deployment, and management of traditional enterprise IT systems through central IT is not sufficient for doing the same with IoT systems. Further, while IoT systems will likely use IT infrastructure, such as wired and wireless networking, deployed and supported by central IT, to support the newly acquired IoT system, it is very likely that central IT will not have the resources or expertise to support the wide-ranging performance aspects required of the IoT system.

IoT systems are unique in that they span many organizations, such as those mentioned above, within an institution. *They are also unique in that they affect many types of risk within an institution* to include financial, reputation, operational, safety, and other types of risk.

For each of the statements or questions below for use in managing vendor relationships, two additional columns are provided: one for type(s) of risk involved and one, e.g., organizations on campus that may be interested in the particular statement or question at hand. In both cases – risk type and organization – it is acknowledged that there can be overlap between types. For example, financial risk can also affect reputation risk. (Almost everything affects an institution's reputation risk.) *The risk item or the organization indicated is primarily intended to be used as examples and potential talking, negotiating, and management points.*

For example, higher education institutional organizations having interest include:

- PI and laboratory staff
- Planning/budgeting office
- Capital development
- Facilities management
- Police department
- Central IT
- Distributed IT groups
- Risk, compliance, Chief Information Security Officer (CISO), and privacy offices.

For example, Higher Education risk areas include:

- Privacy
- Financial
- Operational
- Reputation
- Compliance
- Safety
- Cybersecurity.

Both lists are not exhaustive, and both lists have items that have interdependency on other items. The intention is to consider them in planning, talking, negotiation, and vendor management activities and to inform and elevate the conversation.

Issue/Statement/Question	Example Potential Risk Area	Example Institutional Organization Having Interest
• Does IoT vendor need 1 (or more) data feeds/data sharing from your organization? • Are the data feeds well defined? • Do they exist already? • If not, who will create and support them? • Are there privacy considerations?	Operational, CISO, privacy, ...	Central IT, PI, ...
• How many endpoint devices will be installed? • Is there a patch plan? • Do you do the patching? • Who manages the plan, you or the vendor? • What is involved (labor/time) in a patch in relation to the scale of the IoT system?	Operational, financial, ...	Facilities Mgmt., Central IT, ...
• Does this vendor's system have dependencies on other systems? • If so is that second system (and even subsequent dependencies) changing rapidly? • Is there a plan or resources to manage these interdependency integrations?	Financial, operational, reputation, ...	Central IT, Facilities Mgmt., Capital Dev, ...
• How many IoT systems are you already managing? • How many endpoints do you already have? • Are you anticipating/planning or planning more in the next 18 months?	Financial, operational, reputation, ...	Facilities Mgmt., Central IT, Capital Dev, ...
• Are you following a standard Dev/Test/ Deploy process? Other?	Operational, compliance, ...	Central IT, local IT, Facilities Mgmt., ...
• Is there a commissioning plan?	Financial, compliance, cybersec, ...	Capital Dev, Facilities Mgmt., ...

(Continued)

Issue/Statement/Question	Example Potential Risk Area	Example Institutional Organization Having Interest

• Have IoT vendor deliverable expectations been stated?
  • For example, contract, memorandum of understanding, letter, and others?
  • How does the vendor manage security in the course of delivery?
  • Has the vendor changed default logins and passwords?
  • Has the password schema been shared with you?
  • Are non-required ports closed on all your deployed IoT endpoints?
  • Has the vendor port scanned (or similar) all deployed IoT endpoints after installation?
  • Is there a plan (for you or vendor) to periodically spot check configuration of endpoint devices?
  • Can you find suggestions on how to hack your IOTS from a Google search?
	Operational, financial, compliance, …	Central IT, CISO, Cap Dev, Facilities Mgmt., Planning/budgeting, …

• Has the installed system been documented?
  • Is there (at least) a simple architecture diagram?
  • Server configuration documented?
  • Endpoint IP addresses and ports indicated?
  • Does the documentation follow any sort of standard? Is it readable and consumable across multiple different parties?
	Reputation, operational	Capital Dev, Central IT, Facilities Mgmt., Compliance, …

• Who pays for the vendor's system requirements (e.g., hardware, supporting software, and networking?) • Does local support (staffing/full time equivalent (FTE)) exist to support the installation? Is it available? Will it remain available?	Financial, operational, cybersec	Planning/budgeting, Facilities Mgmt., Central IT, PI/end users, …

(Continued)

Issue/Statement/Question	Example Potential Risk Area	Example Institutional Organization Having Interest
• If supporting IoT servers are hosted in a data center, who pays those costs? • Startup and ongoing costs? • Same for cloud – if hosted in cloud, who pays those costs?Startup and ongoing costs? • Is your approach to cloud hosting based on standard server procedures or on customized services?		

• What is total operational cost after installation?
  • Licensing costs
  • Support contract costs
  • Hosting requirements costs
  • Business resiliency requirements costs
  • Redundancy, recovery, etc. for OS, databases, and apps
	Financial, operational, risk	Facilities Mgmt., Capital Dev, Planning/ budgeting, …

• How can the vendor demonstrate contract performance?
  • Okay to ask vendor to help you figure this out
  • Does the vendor's component include a readily engaged/checked self-test?
	Financial, cybersec	Facilities Mgmt., Capital Dev, Central IT, local IT, …

• Who in your organization will manage the vendor contract for vendor performance? • Without person/team to do this, the contract won't get managed.	Financial, operational, cybersec, …	Planning/ budgeting, CISO, Risk, …
• Can vendor maintenance contract offset local IT support shortages? • If not, then this might not be the deal you want	Financial, operational, …	Facilities Mgmt., Central IT, Cap Dev, …
• For remote support, how does vendor safeguard login and account information? • Do they have a company policy or standard operating procedure that they can share with you?	Cybersec, operational, safety, …	CISO, Central IT, Facilities Mgmt., …

(Continued)

Issue/Statement/Question	Example Potential Risk Area	Example Institutional Organization Having Interest
• In cases where you are administrating access: Does the vendor maintain a back door?	Cybersec, operational, ...	Central IT, CISO, Risk, Compliance, ...
• Is a risk sharing agreement in place between you and the vendor? • Who is liable for what?	Compliance, financial, reputational	CISO, Risk, Facilities Mgmt., Central IT, ...
• What standard of emergency resiliency should the system be built to? • Is there existing emergency power, cooling, if needed?	Operational, reputation, financial, ...	Planning/ budgeting, PI/end user, Risk, ...
• Are there systems for which wireless connections are acceptable? • What criteria should be used to determine if appropriate?	Cybersec, privacy, ...	PI/end user, Central IT, Facilities Mgmt., ...
• How do we include requirements in a Design Guide? • What is the process for updating the guide? • How often? • Are there triggers other than time that lead to a revision?	Operational, financial, compliance, ...	Capital Dev, Facilities Mgmt., Central IT, local IT, ...
• Does the System offer an event monitoring capability?	Operational, cybersec, reputation, ...	Central IT (especially NOC), Facilities Mgmt., CISO, Risk, ...
• Is there a mechanism for integrating System-generated events with institution's existing ticketing system	Operational, ...	Central IT, Facilities Mgmt., local IT, ...

(*Continued*)

Issue/Statement/Question	Example Potential Risk Area	Example Institutional Organization Having Interest
• Does the System offer event trend analysis tools?	Operational, financial, ...	Central IT, Facilities Mgmt., ...
• Does the vendor offer a proposed set of severity and urgency guidelines for Systems events	Operational, ...	Central IT, Facilities Mgmt., ...

Other Resources

■ National Institute of Standards and Technology (NIST) Cybersecurity for IoT Program
  - https://nist.gov/programs-projects/nist-cybersecurity-iot-program
  - http://nvlpubs.nist.gov/nistpubs/SpecialPublications/NIST.SP.800-160.pdf
■ Federal Trade Commission (FTC) & IoT Privacy
  - https://ftc.gov/system/files/documents/reports/federal-trade-commission-staff-report-november-2013-workshop-entitled-internet-things-privacy/150127iotrpt.pdf
■ Industrial Internet of Things Security Framework
  - http://iiconsortium.org/IISF.htm
■ GSMA IoT Security Guidelines
  - http://gsma.com/connectedliving/future-iot-networks/iot-security-guidelines/
■ OWASP IoT Security Guidance
  - https://owasp.org/index.php/IoT_Security_Guidance
■ DHS Strategic Principles for Securing the Internet of Things
  - https://dhs.gov/sites/default/files/publications/Strategic_Principles_for_Securing_the_Internet_of_Things-2016-1115-FINAL pdf
■ Others ...

Potential Future Work

- IoT systems costing
  - Few, if any, institutions have a handle on this
  - Vendor costs, local costs, and total cost of ownership.
- Network segment portfolio strategies
  - Segmentation is a popular concept, but how are those segmentation portfolios managed?
- Internal Industrial Control System (ICS) and IoT exposure
  - Shodan/Censys do public addresses
  - Internal Virtual Local Area Networks, virtual routing and forwarding, etc. not covered.
- Specific checklist for Network Operations Centers (NOCs)
- Benchmark/standard for exposure in higher education (HE).

Appendix B: Making Sense of the Internet of Things in Higher Education[1]

Understanding the potential benefits of Internet of Things (IoT) systems, as well as the potential downfalls stemming from poor implementation and management of these systems, is critical for senior institutional leaders in higher education. The ability to implement and operate these systems well will be a competitive differentiator among other institutions.

IoT systems offer substantial potential for use in postsecondary environments. They can be used to enhance student, faculty, and staff safety through building access systems, provide energy management and conservation capabilities, and enhance learning environments. IoT systems also have broad application in research to include arrays of sensors and actuators that together can meticulously control and monitor research environments. However, IoT systems must be selected, procured, implemented, and managed well in order to see the success of the IoT system investment by the institution.

Not only do IoT systems offer a potential value-add for an institution, they are also essential for the institution to be competitive with its peers and attract the highest caliber of faculty and students possible. This is particularly true in research institutions. Those institutions that do this well – select, procure, implement, and manage IoT systems – will have faculty that are satisfied with their systems. Notably, they will be satisfied with not having to spend their research and teaching time trying to keep their IoT systems running. These faculty will tend to

---

[1] This article has been originally published in *Evolllution Magazine* (https://evolllution.com/technology/infrastructure/making-sense-of-the-internet-of-things-in-higher-education/).

be motivated to remain at that institution – *and* they will tell colleagues about how well things are going.

Universities and colleges that do not implement IoT systems well will tend to have faculty that are less motivated to stay because they are using valuable research and teaching time to keep their IoT systems, which are required for their work, operating well enough to support their respective endeavors. These faculty will have less motivation to stay at that institution – *and they will also* tell colleagues about how not-so-well things are going.

Because institutional experience with IoT systems implementation, management, and support is still nascent yet evolving rapidly, it is easy for senior leadership to have misconceptions – or more commonly, a lack of awareness of some of the pitfalls of these systems when not planned, implemented, and managed well. (It's easier for leadership to see potential benefit of these systems because there are countless vendors knocking down their door or the doors of their respective staffs to say what is possible.)

Below are five aspects of IoT systems that institutional leaders may not be aware of:

1. The T in IoT is a Networked Computer

    The "Thing" in Internet of Things, or the T in IoT, is a device that computes, is networked, and interacts with the environment in some way, such as sensing or moving something. That is, that Thing – regardless of how small, embedded, and unseen it may be – is a *networked computer.* As such, it has the same exposure to malicious online activity as any other networked computer, and often, these Things are deployed by the hundreds, thousands, or more. So, without thoughtful implementation and management, institutions can unwittingly be installing thousands of under-managed or unmanaged networked computing devices ripe for operational failure or malicious compromise.

2. The T in IoT Is a Device That Has Many Different Components

    An IoT device can have many different hardware and software components, and these components may come from many different sources. For example, a device might have a hardware sensor, a piece of firmware that communicates with that hardware, software for networked communication, software for multiple wireless protocols, a web service, business logic, encryption software, and others. Each of these could come from a different source – a large software developer, code developed in somebody's garage, a company in a currently hostile nation-state, or elsewhere. We typically don't know (and typically don't ask) where these components come from or whether they have been vetted (by anybody) – and we may be deploying them by the thousands in our institutions. Because of the scale of deployment, the issue of supply chain awareness, or lack thereof, can be particularly poignant in the IoT environment.

3. IoT systems Support and Management Is Not Free

For successful implementation, in terms of both ROI and cybersecurity, IoT systems need to be appropriately configured and subsequently well managed. This includes all of the devices, possibly in the hundreds or thousands or more, and any supporting server and supporting client applications. Further, the underlying network supporting the IoT system has to be managed and resourced as well. Network segmentation is a popular approach to supporting IoT systems. While this can offer benefits, the network segment also has to be managed.

Institutions will continue to see growth in the number of network segments that must be managed.

4. Central IT Is Probably Not Managing the IoT system

Because IoT systems have so many different routes and purchase paths into an institution, without institutional oversight, it is unlikely that the central IT organization can know about all of the IoT systems. In fact, it may not even know about many of the IoT systems. IoT systems can enter the institution through departmental purchases, minor or major capital purchasing, personal or institutional credit card purchases, or by other means. Once purchased, these systems may only need a Wi-Fi connection or network wall port to begin operation. Central IT may not have been included in IoT system selection – they may not even know that a new IoT system has been purchased and installed. Even if they do know, the central IT group may not have the resources to support the system on its networks.

5. IoT systems Vendors Need Clearly Defined Expectations

Institutions need to be specific in the selection and procurement process regarding exactly what the IoT systems provider will deliver as a part of systems implementation and what they will subsequently manage. For example, if the institution does not explicitly communicate that all installed devices should have default logins and passwords changed and unnecessary services disabled, there is no reason to assume that the vendor will automatically include this as part of the deliverable. Other aspects of the IoT systems deliverable could include network diagram, installation locations, as well as software and firmware version numbers, IP addresses, MAC addresses of all installed devices, patch plan, and other requirements.

These are just a few of the concepts that may not be readily grasped by senior institutional leadership. Indeed, it may not be readily grasped by most. IoT systems are different from traditional enterprise IT systems in many ways, and institutions have very limited to no experience in implementing and managing these systems. Without thoughtful planning and oversight, IoT systems deployments run the risk of delivering a poor return on investment as well as exposing the institution to significantly increased cyber risk.

Appendix C: Interview with Hendrik Van Hemert, Pacific Northwest Regional Director–Technical Services, McKinstry, December 17, 2018

1. Benson: What are red flags regarding likelihood of implementation success when you initially engage with clients (e.g. city, institution, or corporate customer) for implementation of energy management and/or building automation systems (BAS)?

Van Hemert: During new construction or major renovation projects a lack of operational input undermines long-term success. For example, BAS points are named standard names for the vendor but not the client. At [one institution] we see this on every project where sub meters are named things like Panel-A2. That doesn't mean anything to the operational team but they are not asked how to name the sub meter so it makes sense ...

2. Benson: What are indicators that you, as systems provider, leads you to expect that the engagement will be productive and have a high probability of success (and potential for scalability that benefits both the city/institutional customer & provider).

Van Hemert: 1. Broad stakeholder involvement. 2. Creativity in identifying all of the potential use cases of a new system or data set. 3. Prioritized and measurable objectives. When we can land on ~5 KPIs for a project with a client that are long-term

I usually feel good about how the project will go. Example KPIs may be: Ensure CO2 levels in classrooms are under 900 ppm (parts per million) for 95% of the occupied hours. The KPI doesn't state how to get there, but it aligns the project team on the outcome and allows different subject matter experts to come together to solve.

3. Benson: What are the top 2 or 3 challenges that clients new to energy management and building automation systems wrestle with?

Van Hemert: 1. I have seen clients try to solve every problem with initial launch. We think a top-down approach is more effective, specifically, identify overall facility goals, then find the data sets that can help you track performance against those goals, then add analytics that save you time in tracking against those goals. Clients often start with all of their BAS data (50k+ points) and then spend all of their time trying to organize and condition this massive data set. We try to focus on a subset of overall systems and then the points necessary to tell us how it's performing. When we do that engineering review upfront we can often have a very successful program with as little as 100 points of live data. Over time, the program will keep evolving and point lists will increase, but initially, clients get bogged down in too much information and can't make it actionable.

Van Hemert: 2. Implementation is often a problem. Any software system will shoot out recommendations based on faults. But, it often takes a significant amount of work to root cause those faults and then implement. And we are often talking about physical assets so resolving an issue often takes both data analysis and a wrench, few people have both skill sets. Identifying faults but not fixing them just undermines overall program objectives.

4. Benson: Is there a spectrum of readiness/sophistication across potential energy management and BAS clients? Or are most fairly similar?

Van Hemert: Huge spectrum. Usually related to organizational maturity and staff skill sets. By organizational maturity I'm thinking of things like: do they use a CMMS (Computerized Maintenance Management System) or WMS (Work Management System) to manage workflow, do they have procedures for various tasks, do they have good asset tracking, do they have common standards for BAS including things like naming.

5. Benson: Can you comment on the pros & cons of the growing number of building/facility energy & health certifications?

Van Hemert: Bad – confusing to market and not having the signaling power to consumers and buyers like they should. Also, lack of common standard means providers have to learn every one and keep folks certified which adds cost to the overall system. Good – buildings serve very different end uses. Having a standard that fits for all would likely be watered down. Having many standards allows a developer to really define what type of building they want and how they want the building to express their values. And, many clients end up learning significantly by reviewing and then making a decision on which path to go down. Many building developers end up borrowing pieces from multiple standards and even if they go with LEED they may borrow heavily from WELL in their direction to contractors.

6. Benson: What does actionable data mean to you as a systems provider?

Van Hemert: The data must have context. That context usually comes from more static information (place, type, procedure, design intent…). The live data is only actionable when there is contextual data that helps you define thresholds or alarms and then helps you actually resolve. The hard work is in the middle with workflow design and management … live data is only of value if you have foundational or static data to go with it.

Index